Winds in the Atmospheric Boundary Layer

Junyi He · Qiu-Sheng Li

Winds in the Atmospheric Boundary Layer

From Meteorology to Engineering Applications

Springer

Junyi He
Department of Architecture and Civil Engineering
City University of Hong Kong
Hong Kong, Hong Kong

Qiu-Sheng Li
Department of Architecture and Civil Engineering
City University of Hong Kong
Hong Kong, Hong Kong

ISBN 978-981-95-1516-5 ISBN 978-981-95-1517-2 (eBook)
https://doi.org/10.1007/978-981-95-1517-2

This Springer imprint is published by the registered company Springer Nature Singapore Pte Ltd.
The registered company address is: 152 Beach Road, #21-01/04 Gateway East, Singapore 189721, Singapore

Preface

This book describes the structure of winds in the atmospheric boundary layer (ABL), with a special focus on its implications for wind engineering, civil structural design, and wind energy applications.

Several books on the ABL have already been published. Stull's book Introduction to Atmospheric Boundary Layer is probably the earliest and the most famous one. A recent example is Foken's book Micrometeorology. A more theoretical one is Wyngaard's book Turbulence in the Atmosphere.

These books are all written by meteorologists. Although they provided detailed information on the ABL from a fluid dynamics/thermodynamics view, they might not be designed for general audiences with backgrounds other than meteorology (e.g., wind engineering, structural engineering, and wind energy). This book, written by wind engineers, concentrates on the atmospheric wind properties influencing the design and operation of civil structures, such as wind turbines and high-rise buildings. To the authors' knowledge, such a book is currently not available.

Another motivation behind this book lies in the gaps between wind engineers and meteorologists. The knowledge of the ABL meteorology has been drastically enhanced in the last few decades due to the development of measurement instrumentation and simulation technology. Nevertheless, in wind codes and structural design standards, the ABL is still modelled in a "traditional" and simplified manner, mostly based on the studies in the 1960s and 1970s. Some examples of such simplifications include the neglect of atmospheric stability effects, the ignorance of wind direction turning with height, and the oversimple treatment of the ABL height. This book intends to bridge the gaps between wind engineers and meteorologists and introduce the recent advances in the ABL meteorology to the wind engineering community.

This book will report some new observations of winds in the ABL owing to our access to and analysis of a wide variety of meteorological data. The most notable data source comes from the meteorological monitoring network of the Hong Kong Observatory. The measurement instruments include cup anemometers, wind vanes, sonic anemometers, thermometers, barometers, etc., installed at automatic weather stations and buoys, as well as Global Position System (GPS) dropsondes, radiosonde balloons, Doppler radar wind profilers, Light Detection and Ranging (LiDAR) systems, Sonic

Detection and Ranging (SoDAR) systems, and microwave radiometers. Another important database is the 356-m-high Shenzhen Meteorological Tower. The Shenzhen Tower is the tallest of its kind in Asia and the second tallest in the world, with its height exceeding many famous meteorological towers (e.g., those in Cabauw, Hamburg, and Beijing). It has thirteen measurement platforms between heights of 10 m and 350 m, equipped with various sensors to measure the mean and fluctuating components of wind, temperature, humidity, pressure, aerosol concentration, etc. In addition, the observations from the marine platforms in Hong Kong waters facilitate the marine ABL studies presented in this book.

This book will first review the fundamental physics governing the ABL and introduce the sources of meteorological data. Then it will discuss the behaviours of the ABL over both homogenous and complex surfaces. The vertical structure of mean wind speed, as well as turbulence properties, e.g., turbulence intensity, gust factor, turbulence integral length scale, and wind spectra, will be covered in the discussions. The atmospheric stability effects will be explained. Subsequently, this book will examine the wind structure of two special types of windstorms, including tropical cyclones (also known as typhoons or hurricanes) and thunderstorms, which are of main concern for wind-resistant structural design and disaster prevention. The long-term climatology of wind will also be briefly discussed, which is expected to be useful for wind resource assessment and wind energy development.

In view of the increasing number of wind-sensitive civil structures and the rapid expansion of wind energy deployments, the wind engineering community urgently needs more knowledge of the wind structure in the ABL. This book is expected to fulfil this need. This book is expected to help meteorologists to understand the concerns and needs of wind engineers, bring the updated knowledge of the ABL meteorology to wind engineers' attention, and thus enable more cost-efficient design and operation of civil structures against winds.

Hong Kong

Junyi He
Qiu-Sheng Li

Contents

1 Introduction to the Atmospheric Boundary Layer 1
1.1 Definition and Vertical Structure of the Atmospheric Boundary Layer 1
1.2 Categories of Winds in the Atmospheric Boundary Layer: Mean Wind, Turbulence, and Waves 3
1.3 Wind Spectra and Spectral Gap 6
References 7

2 Theory 9
2.1 Basic Equations 9
2.1.1 Governing Equations for Atmospheric Boundary Layer 9
2.1.2 Simplified Mean Flow Equations 12
2.1.3 Turbulence Closure, K Theory, and Mixing Length 13
2.1.4 Turbulent Kinetic Energy Equation 14
2.2 Surface Layer and Ekman Layer 15
2.2.1 Surface Layer 15
2.2.2 Ekman Layer 15
2.3 Atmospheric Stability 16
2.3.1 Virtual Potential Temperature 16
2.3.2 Static Stability 17
2.3.3 Dynamic Stability 18
2.4 Similarity Theory 19
2.4.1 Monin–Obukhov Similarity 19
2.4.2 Rossby Number Similarity 20
2.4.3 Flux-Variance Similarity 21
2.4.4 Spectral Similarity 22
References 23

3 Wind Observation and Modeling Techniques 25
3.1 In Situ Measurements 25
3.1.1 Cup Anemometer 26
3.1.2 Sonic Anemometer 27
3.1.3 Other Instruments for in Situ Wind Measurements 28
3.2 Remote Sensing Observations 29
3.2.1 Radar Wind Profiler 29
3.2.2 Sodar 30
3.2.3 Lidar 31
3.2.4 Satellite 32
3.3 Numerical Models 33
3.3.1 Numerical Weather Prediction 33
3.3.2 Reanalysis 34
3.3.3 Computational Fluid Dynamics 35
References 37

4 Mean Wind Profiles Over Homogenous Terrain 39
4.1 Wind Profiles in Neutral Conditions 39
4.1.1 Power Law 39
4.1.2 Logarithmic Law 40
4.1.3 Wind Profiles Above the Surface Layer 42
4.1.4 Surface Roughness 44
4.2 Wind Profiles in Non-neutral Conditions 48
4.2.1 Wind Profiles in Unstable and Stable Conditions 48
4.2.2 Wind Profiles in Conventionally Neutral Conditions 52
4.2.3 Low-Level Jets 54
4.3 Wind Profiles in Storms 56
4.3.1 Wind Profiles in Tropical Cyclones 56
4.3.2 Wind Profiles in Thunderstorms 61
References 65

5 Turbulence Over Homogenous Terrain 69
5.1 Turbulence in Neutral Conditions 69
5.1.1 Turbulence Intensity 69
5.1.2 Gust Factor 73
5.1.3 Turbulence Integral Length Scale 76
5.1.4 Wind Spectra 79
5.1.5 Coherence 83
5.2 Turbulence in Non-neutral Conditions 85
5.3 Turbulence in Storms 87
5.3.1 Turbulence in Tropical Cyclones 87
5.3.2 Turbulence in Thunderstorms 90
References 95

6 Winds Over Complex Terrain 99
6.1 Roughness Change and Internal Boundary Layer 99
6.2 Topographic Effects 102
6.3 Winds Over Cities 103
6.3.1 Atmospheric Stability Over Urban Surface 104
6.3.2 Wind Profiles in the Urban Boundary Layer 106
6.4 Winds Over Sea 108
6.4.1 Sea Surface Roughness and Wind Profiles 109
6.4.2 Turbulence and Gusts Over Sea 112
References 114

7 Wind Climatology 117
7.1 Probability Distribution of Wind Speed 117
7.1.1 Weibull Distribution 117
7.1.2 Other Unimodal Distributions 120
7.1.3 Bimodal Distributions 121
7.2 Long-Term Mean Wind Profiles 123
7.2.1 Vertical Profiles of Long-Term Mean Wind Speed 123
7.2.2 Vertical Profiles of Weibull Shape Parameter 125
7.3 Macroscale Wind Spectra 126
7.4 Climate Change 128
7.4.1 Global Climate Models 128
7.4.2 Statistical Downscaling 130
7.4.3 Dynamical Downscaling 132
7.4.4 Climate Change Impacts on Winds 133
References 134

8 Outlook 137

Chapter 1
Introduction to the Atmospheric Boundary Layer

This chapter introduces the basics of the atmospheric boundary layer. Definitions and vertical structures of the atmospheric boundary layer are first presented (Sect. 1.1), followed by a discussion on mean winds, turbulence, and waves (Sect. 1.2). Subsequently, a brief description of the wind spectrum (Sect. 1.3) is provided. This chapter focuses on the meteorological basics, which are then expanded in Chap. 2 on theoretical aspects and Chap. 3 on experimental and numerical methods.

1.1 Definition and Vertical Structure of the Atmospheric Boundary Layer

The Earth's surface is the lowest boundary of the atmosphere. The momentum, heat, and mass transport processes at this surface influence the lowest part (generally from several tens of meters to several kilometers) of the atmosphere, forming the atmospheric boundary layer (ABL). According to Stull (1988), the ABL can be defined as "*the boundary layer as that part of the troposphere that is directly influenced by the presence of the earth's surface, and responds to surface forcings with a timescale of about an hour or less*". Examples of these forcings include frictional drag, heat transfer, evaporation and transpiration, pollutant emission, and terrain-induced flow. Indirectly, the rest of the air above the ABL, which is known as free atmosphere, may be influenced by the surface properties, but this response is rather slow (Stull, 1988).

The vertical structure of a typical ABL is shown in Fig. 1.1. The bottom 10% of the ABL is called the surface layer (also referred to as Prandtl layer or constant flux layer), where the turbulent fluxes vary by no more than 10%. The surface layer can be further divided into three sublayers, including the laminar sublayer, viscous sublayer, and inertial sublayer. In the lowest few millimeters of the surface layer, there is a very thin layer called the laminar sublayer, where molecular transport dominates.

J. He and Q.-S. Li, *Winds in the Atmospheric Boundary Layer*,
https://doi.org/10.1007/978-981-95-1517-2_1

Above this laminar sublayer, the viscous sublayer (or buffer layer) extends for a few centimeters, characterized by mixed exchange conditions. Above the viscous sublayer is the inertial sublayer, which is generally characterized by a sufficiently large Reynolds number (in the order of 10^7–10^8), turbulent flow, and a logarithmic wind profile (see Chap. 4). Immediately above tall vegetation or buildings up to about twice the canopy height, there is an additional sublayer called the roughness sublayer. In the roughness sublayer, turbulent mixing is enhanced, and the fluxes are not constant. Above the surface layer, the remaining 90% of the ABL is the Ekman layer (upper layer), where the fluxes vary with height, the Coriolis force plays an important role in regulating the flow, and the changes of wind direction take place. See Sect. 2.2 for equations governing the surface layer and Ekman layer. At the top of the ABL typically lies a stable inversion layer, where turbulence is intermittent, and the free atmosphere exchanges momentum, heat, and mass with the ABL. Above the ABL and in the free atmosphere, the flow is usually much less turbulent, and the Coriolis force and the pressure gradient force are generally in approximate balance (see Sect. 2.1.2). The vertical structure of the ABL can change with respect to wind strength, terrain roughness, Coriolis parameter, and atmospheric stratification, and the height of the ABL may vary from as low as 30 m in very stable conditions to up to 3 km in highly convective conditions (Bolton & Hakim, 2013). These will be discussed in detail in later chapters.

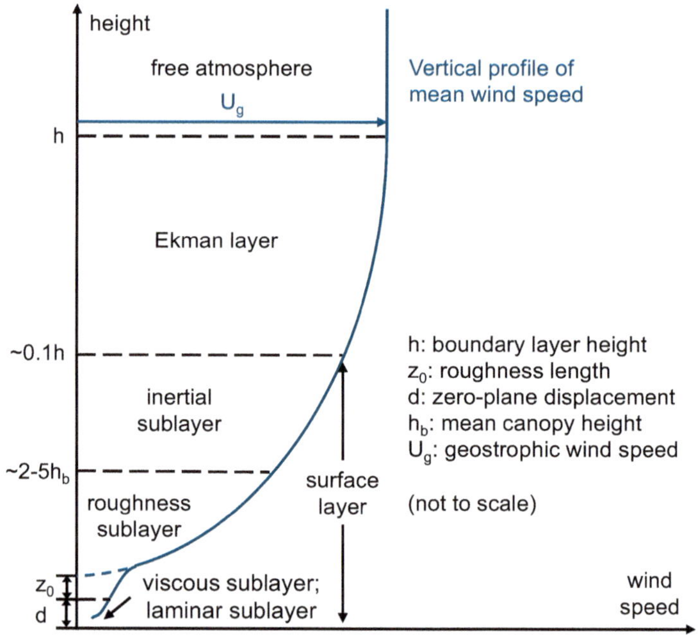

Fig. 1.1 Vertical structure of the atmospheric boundary layer

1.2 Categories of Winds in the Atmospheric Boundary Layer: Mean Wind, Turbulence, and Waves

Winds in the ABL can be classified into three main categories, including mean wind, turbulence, and waves. A common approach to studying them involves separating variables, such as wind speed, into a mean component and a perturbation component:

$$U = \overline{U} + U' \tag{1.1}$$

where U is the total wind speed, $\overline{U}$ is the mean component of the wind speed, and U' is the perturbation component of the wind speed. An example of such separation is shown in Fig. 1.2.

The mean component $\overline{U}$ reflects the effects of the mean wind, while the perturbation component U' represents either turbulence effects or wave effects superimposed on the mean wind. Horizontal mean wind (Fig. 1.3a) has typical speeds of 0–10 m/s in normal circumstances and extreme speeds up to 100 m/s in severe windstorms (such as tropical cyclones and tornadoes). It facilitates rapid horizontal transport (i.e., advection) of momentum, heat, and other scalars such as humidity and pollutants in the atmosphere. Due to frictional effects, wind speeds are generally lower close to the surface, and this change of wind speed with height is known as vertical wind shear. Compared to horizontal mean wind speed, vertical mean wind speeds are considerably less, typically within a few cm/s.

Turbulence almost always exists in the ABL due to the continuous turbulence generation by the strong vertical wind shear near the surface. As shown in Fig. 1.3b, turbulence is characterized by an irregular and quasi-random pattern, which distinguishes it from other types of motion, such as waves. Turbulence primarily influences the vertical transport of momentum, heat, moisture, and pollutants. Outside the ABL,

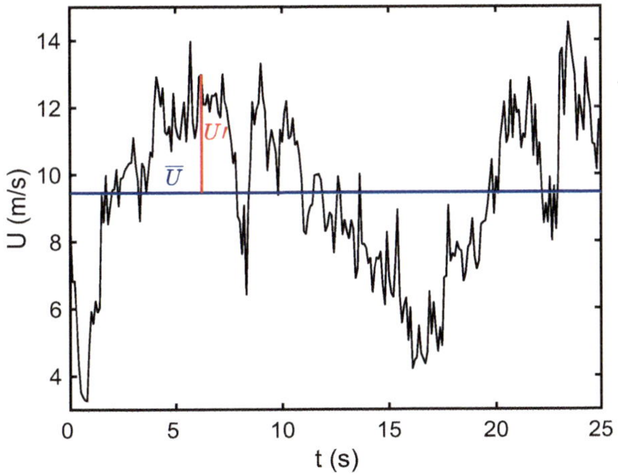

Fig. 1.2 Wind speed observed in a strong wind event, showing U' as the deviation of the actual instantaneous wind U from the mean $\overline{U}$

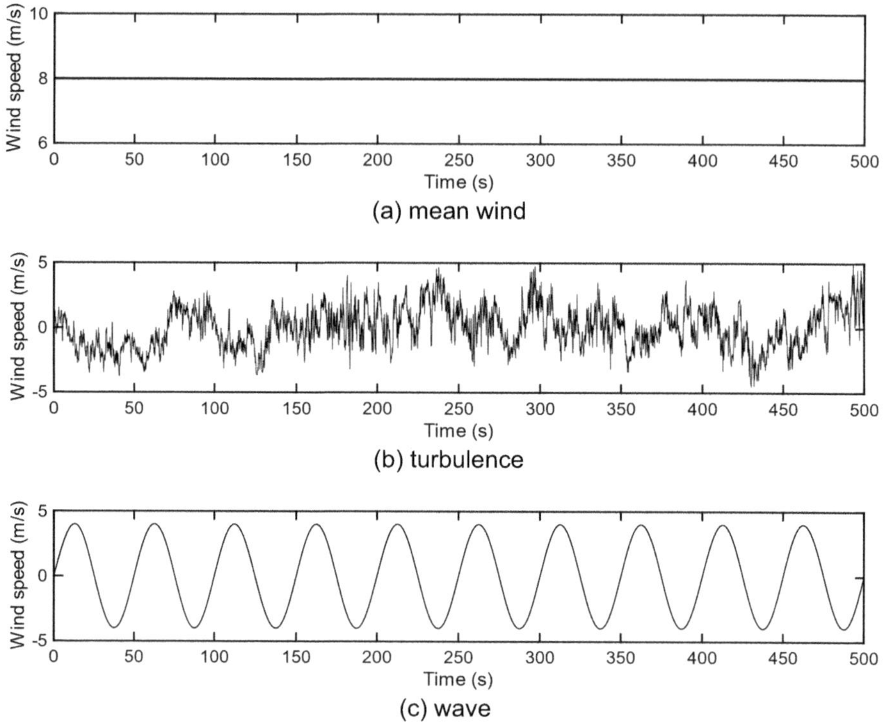

Fig. 1.3 Illustrative diagram of **a** mean wind, **b** turbulence, and **c** wave

turbulence primarily occurs within convective clouds or near the jet stream, where significant wind shears can lead to clear air turbulence.

Waves are commonly observed in the nighttime ABL. As illustrated in Fig. 1.3c, waves exhibit a periodic pattern that is much more "regular" than turbulence. While waves are quite effective in transporting momentum and energy, they play a minimal role in transporting heat, humidity, and other substances like pollutants. Waves can arise locally due to shear in the mean wind or from the flow over obstacles such as mountains. They can also propagate from distant sources, including thunderstorms or explosions. Atmospheric waves may sometimes amplify wind shears in specific areas, resulting in the development of turbulence.

Microscale turbulence is a three-dimensional phenomenon. Consequently, turbulence in the along-wind direction is often accompanied by turbulence in the crosswind and vertical directions. Moreover, turbulent motions can also lead to changes in temperature and humidity when there is a gradient of these variables. Therefore, each of these variables can be partitioned into mean and turbulent components:

$$u = \overline{u} + u' \tag{1.2}$$

$$v = \overline{v} + v' \tag{1.3}$$

$$w = \overline{w} + w' \tag{1.4}$$

$$T = \overline{T} + T' \tag{1.5}$$

$$q = \overline{q} + q' \tag{1.6}$$

where u, v, and w are the wind velocity components in the x, y, and z directions in the three-dimensional space, T is the air temperature, and q is the specific humidity.

The splitting technique in these equations can be applied to the equations of motion, resulting in several new terms (to be discussed in Sect. 2.1). Some of these terms, which consist of products of perturbation variables, describe nonlinear interactions associated with turbulence. When the focus is on wave motions, these terms are typically neglected. Conversely, when the emphasis is on turbulence, other terms that contain only a single perturbation variable describe linear motions linked to waves, and these are disregarded.

The wind velocity components u, v, and w are interrelated, exhibiting nonzero correlation products that highlight their dependence on one another. Among these correlations, the most significant one is the surface momentum flux:

$$\overline{u'w'} = \frac{1}{T}\int_0^T u'(t)w'(t)dt \tag{1.7}$$

The overbar represents the time average over the period T, and the prime denotes the deviation from the average. The surface momentum flux is typically negative, and this negativity arises from the fact that mean wind speed tends to increase with altitude in the surface layer. Specifically, downward fluctuations in the vertical wind component w' tend to pull positive longitudinal fluctuations u' from greater heights downwards, while upward fluctuations in w' are associated with negative u' from lower heights. In addition to the surface momentum flux, there are other important correlation products, such as the surface heat flux $\overline{w'\theta'}$ (where θ is the potential temperature) and surface moisture flux $\overline{w'q'}$ (Foken, 2017).

The square root of the magnitude of the surface momentum flux is referred to as friction velocity:

$$u_* = \sqrt{\left|\overline{u'w'}\right|} \tag{1.8}$$

This parameter is commonly used as a velocity scale in the ABL. It serves as an indicator of how effectively horizontal momentum is transferred downward by turbulent motions within this layer.

1.3 Wind Spectra and Spectral Gap

The turbulent flow can be thought of as a superposition of eddies, which are irregular swirls of motion. Most turbulence in the ABL is generated by forces originating from the ground. For instance, solar heating of the surface leads to the formation of rising columns of warmer air, which can be understood as large eddies. Surface drag on wind flow over the ground creates wind shear, which often develops into turbulent conditions. Mountains and buildings disrupt the airflow, resulting in turbulent wakes downwind of these structures.

The largest eddies scale to the ABL height (from several tens of meters to several kilometers), which are generated directly by the surface forcings and are thus generally the most intense. Large eddy motions can generate smaller and weaker eddies, and the energy transfer process from the large to small eddies is referred to as the energy cascade. At the Kolmogorov microscale (usually about 1 mm), the molecular viscosity dissipates the energy into heat (Kaimal & Finnigan, 1994). The relative strengths of the different scale eddies define the turbulence spectrum. By conducting a spectrum analysis of a turbulent signal using mathematical methods, we can derive the turbulence spectrum and assess the contribution of different sizes of eddies to the turbulent kinetic energy.

A typical spectrum of wind speed measured near the ground is shown in Fig. 1.4 (Van der Hoven, 1957). The vertical axis represents the turbulence energy distribution across various eddy sizes, while the horizontal axis reflects the size of these eddies in terms of their periods and frequency of wind speed fluctuations. Smaller eddies have shorter periods than their larger counterparts. The peaks in the spectrum highlight the sizes of eddies that contribute most significantly to turbulent kinetic energy. The peak at the left side, associated with a period of approximately four days, corresponds to wind speed variations linked to weather fronts and weather systems, such as the Rossby-wave cycle. The subsequent peak, occurring at a one-day interval, is indicative of the diurnal wind speed cycle. Finally, the right peak represents microscale eddies, characterized by durations ranging from 30 s to 3 min.

There is a notable absence of wind speed variations within the period of approximately 10 min–2 h, which is called the spectral gap. Periods longer than this gap are related to the synoptic scale mean flow, while those shorter than this gap are linked to the microscale turbulence. The existence of the spectral gap supports the notion of fast fluctuations as stationary processes, which allows for the analysis of time series using statistically stationary theories and models (Larsén et al., 2016). By averaging the wind speeds over a period located in the spectral gap, it becomes possible to distinguish between the mean flow and turbulent components, as illustrated in Fig. 1.2 and described in Eqs. (1.1)–(1.6). The turbulent part of the spectrum, which is important to the design of civil structures, will be discussed in detail in Sect. 5.1.4.

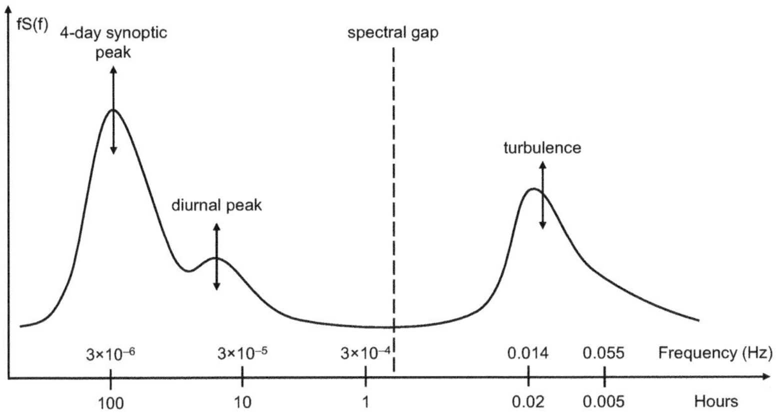

Fig. 1.4 Schematic spectrum of wind speed

References

Bolton, J. R., & Hakim, G. J. (2013). *An introduction to dynamic meteorology*. Elsevier Academic Press.

Foken, T. (2017). *Micrometeorology* (2nd ed.). Springer.

Kaimal, J. C., & Finnigan, J. J. (1994). *Atmospheric boundary layer flows: Their structure and measurement*. Oxford University Press.

Larsén, X. G., Larsen, S. E., & Petersen, E. L. (2016). Full-scale spectrum of boundary-layer winds. *Boundary-Layer Meteorology, 159*, 349–371. https://doi.org/10.1007/s10546-016-0129-x

Stull, R. B. (1988). *An introduction to boundary layer meteorology* (1st ed.). Springer.

Van der Hoven, I. (1957). Power spectrum of horizontal wind speed in the frequency range from 0.0007 to 900 cycles per hour. *Journal of Meteorology, 14*, 160–164. https://doi.org/10.1175/1520-0469(1957)014%3c0160:PSOHWS%3e2.0.CO;2

Chapter 2
Theory

This chapter introduces the basic theories that govern the fluid mechanics in the atmospheric boundary layer, including the basic equations (Sect. 2.1) and surface layer and Ekman layer theories (Sect. 2.2). However, in most circumstances, it is insufficient to derive mathematical models for the variables of interest (e.g., mean wind profile) based only on the theories and equations given in Sects. 2.1 and 2.2. Alternatively, atmospheric stability concepts (Sect. 2.3) and similarity theory (Sect. 2.4) provide a way for constructing such models based on observations. The similarity relationships serve as the foundation of modern boundary-layer meteorology.

For those who are interested only in applications, such as design of civil structures against wind, wind energy resource assessment, and pollutant dispersion modeling, they may choose to skip this chapter and directly go to Chap. 3.

2.1 Basic Equations

2.1.1 Governing Equations for Atmospheric Boundary Layer

The governing equations of ABL can be formulated using the theory of homogeneous incompressible fluids considering that the mean and fluctuating air density changes usually no more than 10% in the lowest several kilometers of the atmosphere. The momentum equation (i.e., Navior–Stokes equation) and continuity equation for incompressible flow with constant viscosity in vector form are given as (for the derivation of these equations, please refer to fluid mechanics textbooks such as Çengel and Cimbala (2018)):

Momentum equation:

$$\rho \frac{D\mathbf{U}}{Dt} = -\nabla p + \rho \mathbf{g} + \mu \nabla^2 \mathbf{U} \tag{2.1}$$

J. He and Q.-S. Li, *Winds in the Atmospheric Boundary Layer*,
https://doi.org/10.1007/978-981-95-1517-2_2

Continuity equation:

$$\nabla \cdot \mathbf{U} = 0 \tag{2.2}$$

where ρ is the density of the fluid, $\mathbf{U}$ is the velocity vector, t is the time, ∇ is the divergence, p is the pressure, $\mathbf{g}$ is the gravity vector, and μ is the dynamic viscosity. Note that the vector equation Eq. (2.1) consists of three equations for the three velocity components in three-dimensional space.

Notwithstanding the incompressible approximation, in the real ABL, the variation in air density cannot be completely neglected as they are needed for calculating the buoyancy force. Therefore, the Boussinesq approximation (i.e., density is regarded constant except in the buoyancy term) is adopted for this situation. Moreover, the use of the non-inertial geocentric reference frame (i.e., a frame that is at rest relative to the rotating Earth) in describing the atmospheric motions necessitates the consideration of the Coriolis force arising from the coordinate acceleration. In addition, the ABL thermodynamics should be taken into account with reference to the potential temperature field. In view of the above, the governing equations of the ABL in Cartesian coordinates are expressed as (Bolton & Hakim, 2013):

x-Momentum equation:

$$\frac{Du}{Dt} = -\frac{1}{\rho}\frac{\partial p}{\partial x} + fv + \mu\nabla^2 u \tag{2.3}$$

y-Momentum equation:

$$\frac{Dv}{Dt} = -\frac{1}{\rho}\frac{\partial p}{\partial y} - fu + \mu\nabla^2 v \tag{2.4}$$

z-Momentum equation:

$$\frac{Dw}{Dt} = -\frac{1}{\rho}\frac{\partial p}{\partial z} + g\frac{\theta}{\theta_0} + \mu\nabla^2 w \tag{2.5}$$

Continuity equation:

$$\frac{\partial u}{\partial x} + \frac{\partial v}{\partial y} + \frac{\partial w}{\partial z} = 0 \tag{2.6}$$

Thermodynamic energy equation:

$$\frac{D\theta}{Dt} = -w\frac{\partial \theta_0}{\partial z} \tag{2.7}$$

where $D/Dt = \partial/\partial t + u\partial/\partial x + v\partial/\partial y + w\partial/\partial z$, u, v, and w are the velocity components in the x, y, and z directions usually taken as zonal, meridional, and vertical directions, respectively (i.e., $\mathbf{U} = \mathbf{i}u + \mathbf{j}v + \mathbf{k}w$, where $\mathbf{i}$, $\mathbf{j}$, and $\mathbf{k}$ are unit vectors in x, y, and z directions, respectively), f is the Coriolis parameter and $f = 2\Omega \sin\varphi$ (where Ω is the rotation rate of Earth and φ is the latitude), θ_0 is the basic state value of potential temperature, and θ is the departure of potential temperature from θ_0. In moist flows, θ and θ_0 should be replaced by the virtual potential temperature θ_v and θ_{v0} (see Sect. 2.3.1). From left to right, the terms of the x- and y-momentum equations represent the tendency and advection (term I), the pressure gradient force (term II), the Coriolis force (term III), and the viscous stress (term IV). In the z-momentum equation, the Coriolis force term is replaced by the buoyancy term.

To account for the turbulent flow, all the variables in Eqs. (2.3)–(2.7) need to be decomposed into mean and fluctuating components, i.e., $u = \overline{u} + u'$, $v = \overline{v} + v'$, $w = \overline{w} + w'$, $\theta = \overline{\theta} + \theta'$. The governing equations for mean motions thus have the form:

x-Momentum equation:

$$\frac{D\overline{u}}{Dt} = -\frac{1}{\rho}\frac{\partial \overline{p}}{\partial x} + f\overline{v} - \left[\frac{\partial \overline{u'u'}}{\partial x} + \frac{\partial \overline{u'v'}}{\partial y} + \frac{\partial \overline{u'w'}}{\partial z}\right] + \mu\nabla^2\overline{u} \tag{2.8}$$

y-Momentum equation:

$$\frac{D\overline{v}}{Dt} = -\frac{1}{\rho}\frac{\partial \overline{p}}{\partial y} - f\overline{u} - \left[\frac{\partial \overline{u'v'}}{\partial x} + \frac{\partial \overline{v'v'}}{\partial y} + \frac{\partial \overline{v'w'}}{\partial z}\right] + \mu\nabla^2\overline{v} \tag{2.9}$$

z-Momentum equation:

$$\frac{D\overline{w}}{Dt} = -\frac{1}{\rho}\frac{\partial \overline{p}}{\partial z} + g\frac{\overline{\theta}}{\theta_0} - \left[\frac{\partial \overline{u'w'}}{\partial x} + \frac{\partial \overline{v'w'}}{\partial y} + \frac{\partial \overline{w'w'}}{\partial z}\right] + \mu\nabla^2\overline{w} \tag{2.10}$$

Continuity equation:

$$\frac{\partial \overline{u}}{\partial x} + \frac{\partial \overline{v}}{\partial y} + \frac{\partial \overline{w}}{\partial z} = 0 \tag{2.11}$$

Thermodynamic energy equation:

$$\frac{D\overline{\theta}}{Dt} = -\overline{w}\frac{\partial \theta_0}{\partial z} - \left[\frac{\partial \overline{u'\theta'}}{\partial x} + \frac{\partial \overline{v'\theta'}}{\partial y} + \frac{\partial \overline{w'\theta'}}{\partial z}\right] \tag{2.12}$$

where the "additional" terms in the bracket represent the turbulent flux divergence. Even when only the mean velocity components are of concern, these turbulent flux terms cannot be neglected since they are of the same order as the other terms in Eqs. (2.8)–(2.12). Thus, alongside the five unknown mean variables ($\overline{u}$, $\overline{v}$, $\overline{w}$, $\overline{p}$, $\overline{\theta}$) there are additional nine unknown turbulent flux terms (e.g., $\overline{u'w'}$ and $\overline{w'\theta'}$), and the number of unknowns is larger than the number of equations. As a result, the complete equations for the mean flow, unlike the governing equations for the total flow, are not a closed set. This so-called turbulence closure problem remains an unsolved problem (Stull, 1988). To solve the equations, the unknown turbulent flux terms need to be approximated by the five known mean variables, which will be discussed in Sect. 2.1.3.

2.1.2 Simplified Mean Flow Equations

The mean flow equations can be simplified under certain circumstances. For regions without horizontal inhomogeneities (e.g., away from shorelines, towns, and forest edges), the horizontal derivative terms (i.e., terms with $\partial/\partial x$ or $\partial/\partial y$) can be neglected. In addition, at heights well above the viscous sublayer, molecular viscosity terms (i.e., $\mu\nabla^2\overline{u}$, $\mu\nabla^2\overline{v}$, and $\mu\nabla^2\overline{w}$) can be excluded. The horizontal momentum equations for the mean flow are then simplified as:

x-Momentum equation:

$$\frac{D\overline{u}}{Dt} = -\frac{1}{\rho}\frac{\partial\overline{p}}{\partial x} + f\overline{v} - \frac{\partial\overline{u'w'}}{\partial z} \tag{2.13}$$

y-Momentum equation:

$$\frac{D\overline{v}}{Dt} = -\frac{1}{\rho}\frac{\partial\overline{p}}{\partial y} - f\overline{u} - \frac{\partial\overline{v'w'}}{\partial z} \tag{2.14}$$

Moreover, for synoptic scale motions without isobar curvature, the inertial acceleration terms (i.e., $D\overline{u}/Dt$ and $D\overline{v}/Dt$) can be neglected. Thus, to a first approximation, governing equations for mean motions in the ABL express a three-force balance among the pressure gradient force, the Coriolis force, and the turbulent momentum flux divergence, as follows:

x-Momentum equation:

$$-\frac{1}{\rho}\frac{\partial\overline{p}}{\partial x} + f\overline{v} - \frac{\partial\overline{u'w'}}{\partial z} = 0 \tag{2.15}$$

y-Momentum equation:

$$-\frac{1}{\rho}\frac{\partial\overline{p}}{\partial y} - f\overline{u} - \frac{\partial\overline{v'w'}}{\partial z} = 0 \tag{2.16}$$

Furthermore, outside the ABL, the turbulent momentum flux divergence can be neglected, then the pressure gradient force and the Coriolis force are in approximately balance. The governing equations are further simplified as:

x-Momentum equation:

$$-\frac{1}{\rho}\frac{\partial \overline{p}}{\partial x}+f\overline{v}=0 \tag{2.17}$$

y-Momentum equation:

$$-\frac{1}{\rho}\frac{\partial \overline{p}}{\partial y}-f\overline{u}=0 \tag{2.18}$$

which is called the geostrophic approximation. Under this approximation, a horizontal wind velocity field $\mathbf{U_g} = \mathbf{i}u_g + \mathbf{j}v_g$, called the geostrophic wind, can be defined as:

$$\mathbf{U_g} = \mathbf{k} \times \frac{1}{\rho f}\nabla p \tag{2.19}$$

i.e.,

$$u_g = -\frac{1}{\rho f}\frac{\partial \overline{p}}{\partial x};\ v_g = \frac{1}{\rho f}\frac{\partial \overline{p}}{\partial x} \tag{2.20}$$

which is a good approximate solution for the horizontal wind velocity field above the ABL for synoptic scale weather systems.

2.1.3 Turbulence Closure, K Theory, and Mixing Length

As discussed in Sect. 2.1.1, the governing equations for the turbulent flow are not closed. Thus, the turbulent flux divergence terms (e.g., $\partial\overline{u'w'}/\partial z$, $\partial\overline{v'w'}/\partial z$, and $\partial\overline{w'\theta'}/\partial z$) need to be parameterized as a function of the known quantities so as to close the equations. A simple but effective approach to approximate these terms is to treat turbulent eddies similarly to molecular diffusion such that the flux is proportional to the local gradient of the mean, as follows:

$$\overline{u'w'} = -K_m\left(\frac{\partial \overline{u}}{\partial z}\right);\ \overline{v'w'} = -K_m\left(\frac{\partial \overline{v}}{\partial z}\right);\ \overline{w'\theta'} = -K_h\left(\frac{\partial \overline{\theta}}{\partial z}\right) \tag{2.21}$$

where K_m and K_h are the vertical eddy diffusivity (or eddy viscosity) for momentum and heat, respectively. This closure scheme is known as K theory.

To determine a suitable model for K, the simplest approach is based on Prandtl's mixing length hypothesis (Prandtl, 1925). The hypothesis assumes that a parcel of fluid displaced vertically for a characteristic distance ξ' will mix with the environment and generate a turbulent fluctuation. The magnitude of this fluctuation is determined by ξ' and the vertical gradient of the mean variable:

$$u' = -\xi'\frac{\partial \overline{u}}{\partial z};\ v' = -\xi'\frac{\partial \overline{v}}{\partial z};\ \theta' = -\xi'\frac{\partial \overline{\theta}}{\partial z} \tag{2.22}$$

Considering neutral stability where the buoyancy effects are negligible, the horizontal scale of the eddies is expected to be proportional to their vertical scale so that $w' = c\xi'|\partial \mathbf{U}/\partial z|$, where c is a constant for proportionality, and $\mathbf{U} = \mathbf{i}\overline{u} + \mathbf{j}\overline{v}$ is the horizontal mean velocity vector. Then the vertical turbulent fluxes can be written as (taking u-momentum as an example):

$$\overline{u'w'} = c\overline{\xi'^2}\left|\frac{\partial \mathbf{U}}{\partial z}\right|\frac{\partial \overline{u}}{\partial z} \tag{2.23}$$

then the vertical eddy diffusivity can be expressed as $K_m = c\overline{\xi'^2}|\partial \mathbf{U}/\partial z| = \overline{l^2}|\partial \mathbf{U}/\partial z|$, where l is the mixing length given by $l = \left(c\overline{\xi'^2}\right)^{1/2}$.

2.1.4 *Turbulent Kinetic Energy Equation*

Turbulence is dissipative by viscous diffusion. Thus, the ABL can be turbulent only if there are physical processes that continuously produce turbulence. To investigate this, the mean equations for velocity components (Eqs. 2.8–2.10) are firstly subtracted from the corresponding unaveraged equations (Eqs. 2.3–2.5), then multiplied by u', v', and w', respectively. After averaging the resulting equations, the turbulent kinetic energy (TKE) equation is given as:

$$\frac{Dk}{Dt} = -\overline{u'w'}\frac{\partial \overline{u}}{\partial z} - \overline{v'w'}\frac{\partial \overline{v}}{\partial z} + \frac{g}{\theta_{v0}}\overline{w'\theta_v'} - \frac{\partial \overline{w'k}}{\partial z} - \frac{1}{\rho}\frac{\partial \overline{w'p'}}{\partial z} - \varepsilon \tag{2.24}$$

where $k = \left(\overline{u'^2} + \overline{v'^2} + \overline{w'^2}\right)/2$ is the TKE per unit mass, which is a quantitative measure of the turbulent energy contained in the flow, and ε is the rate of TKE dissipation. From left to right, the terms represent the tendency and advection (term I), the shear production (terms II and III), the buoyancy production or consumption (term IV), the TKE transport (term V), the pressure transport (term VI), and the dissipation (term VII).

2.2 Surface Layer and Ekman Layer

2.2.1 Surface Layer

Surface layer is the bottom 10% of the ABL where the turbulent fluxes vary by no more than 10%. In the neutral surface layer, since the eddy size is limited due to the surface, it can be postulated that the mixing length is proportional to the distance to the surface by $l = kz$, where k is a constant of proportionality known as the von Kármán constant. k was reported to vary between 0.35 and 0.42 but is now mostly taken as 0.4 (Foken, 2017). Substituting $K_m = \overline{l^2}|\partial\mathbf{U}/\partial z|$ and $l = kz$ into Eq. (2.21) and defining a characteristic velocity scale (i.e., friction velocity) as $u_* = \left(\overline{u'w'}^2 + \overline{v'w'}^2\right)^{1/4}$ gives:

$$\left|\frac{\partial\mathbf{U}}{\partial z}\right| = \frac{u_*}{kz} \tag{2.25}$$

Assuming that the vertical variation in friction velocity is negligible in the surface layer, integrating Eq. (2.25) with respect to z yields the logarithmic law:

$$U = \frac{u_*}{k}\ln\left(\frac{z}{z_0}\right) \tag{2.26}$$

where $U = \sqrt{\overline{u}^2 + \overline{v}^2}$ is the magnitude of horizontal mean wind velocity, and z_0, known as the roughness length, is a constant of integration chosen so that $U = 0$ at $z = z_0$ (Stull, 1988). The logarithmic law and its variant for non-neutral atmospheric stability will be further discussed in Chap. 4.

2.2.2 Ekman Layer

Ekman layer is the layer above the surface layer and below the ABL top. It comprises approximately 90% of the ABL depth. Substituting the first-order closure scheme in Eq. (2.21) into the governing equations of Eqs. (2.15) and (2.16), rewriting the pressure gradient force in terms of the geostrophic wind in Eq. (2.20), and assuming that the value of K_m is constant, the classical Ekman layer equations are obtained as:

$$f\left(\overline{v} - v_g\right) + K_m\left(\frac{\partial^2\overline{u}}{\partial z^2}\right) = 0 \tag{2.27}$$

$$-f\left(\overline{u} - u_g\right) + K_m\left(\frac{\partial^2\overline{v}}{\partial z^2}\right) = 0 \tag{2.28}$$

Assuming that the boundary conditions are $\overline{u} = 0$ and $\overline{v} = 0$ at $z = 0$, and $\overline{u} \rightarrow u_g$ and $\overline{v} \rightarrow v_g$ at $z \rightarrow \infty$, and the geostrophic wind u_g and v_g are independent of height (i.e., barotropic), the solution for the Northern Hemisphere ($f > 0$) is:

$$\overline{u} = u_g\left[1 - \exp(-\gamma z)\cos\gamma z\right] - v_g \exp(-\gamma z)\sin\gamma z \quad (2.29)$$

$$\overline{v} = v_g\left[1 - \exp(-\gamma z)\sin\gamma z\right] + u_g \exp(-\gamma z)\sin\gamma z \quad (2.30)$$

where $\gamma = \sqrt{f/2K_m}$. This solution is the famous Ekman spiral, which gives the angle between the surface and the geostrophic wind direction to be 45°. The ideal Ekman spiral is seldom observed in the ABL, primarily because momentum fluxes are not generally directly proportional to the mean wind speed gradient. Even if the K theory holds true, it is inappropriate to assume a constant K_m, because K_m should change significantly with height near the surface.

2.3 Atmospheric Stability

The balance of the terms in the TKE equation (Eq. 2.24) influences the flow's capacity to sustain or transition into turbulence, thereby serving as an indicator of flow stability. The following sections will introduce the concept of virtual potential temperature and atmospheric stability, as well as some nondimensional numbers that characterize the stability.

2.3.1 *Virtual Potential Temperature*

Buoyancy plays a key role in driving turbulence within the ABL. Warm air thermals tend to rise (positively buoyant) as they are less dense than the surrounding air. In studies of atmospheric dynamics, virtual temperature is a commonly used variable. It represents the temperature that dry air needs to attain to match the density of moist air at the same pressure. Consequently, researchers often analyze variations in virtual temperature rather than density.

Unsaturated moist air is more buoyant than dry air at the same temperature due to lower density of water vapor. As a result, the virtual temperature of moist, unsaturated air is always higher than its absolute temperature. For unsaturated air with a mixing ratio q, the virtual potential temperature θ_v can be expressed as follows:

$$\theta_v = \theta(1 + 0.61q) \quad (2.31)$$

where θ is the potential temperature, which is used to remove the temperature variation resulting from changes in pressure altitude. The potential temperature θ at pressure P is the temperature an air parcel would achieve if adiabatically adjusted to a standard reference pressure P_0, typically set at 1000 hPa. It is defined as:

$$\theta = T_0\left(\frac{P}{P_0}\right)^{R_d/c_p} \tag{2.32}$$

where T_0 is the absolute air temperature, P is the air pressure, P_0 is the standard reference pressure, R_d is the specific gas constant of air, and c_p is the specific heat capacity of air at constant pressure. $R_d/c_p = 0.286$ is commonly used.

However, liquid water is denser than dry air, which makes saturated air heavier compared to unsaturated air. The presence of water droplets within air is referred to as liquid water loading, and this loading decreases the virtual temperature. The virtual potential temperature for saturated air is given as:

$$\theta_v = \theta(1 + 0.61q_{sat} - q_l) \tag{2.33}$$

where q_{sat} represents the water vapor saturation mixing ratio, and q_l is the liquid water mixing ratio.

2.3.2 *Static Stability*

In general, the term atmospheric stability refers to static stability. Static stability reflects the capability of the air to become turbulent or laminar due to buoyant convection. It is not related to winds but depends on the temperature field. The traditional definition of static stability is based on the lapse rate of virtual potential temperature $\Gamma_{\theta_v} = -d\overline{\theta_v}/dz$. If $\Gamma_{\theta_v} < 0$ so that $\overline{\theta_v}$ increases with height, an air parcel that is displaced adiabatically from its equilibrium level will experience positive buoyancy when moved downward and negative buoyancy when moved upward. As a result, the air parcel will tend to return to its equilibrium level, and the atmosphere is said to be statically stable. On the other hand, if $\Gamma_{\theta_v} > 0$, the displacement of the parcel will increase exponentially in time, and the atmosphere is said to be statically unstable. In the case of $\Gamma_{\theta_v} = 0$, no accelerating force will exist and the parcel will be in neutral equilibrium at its new level, and the atmosphere is said to be statically neutral.

The above definition of static stability is a local definition, which means the stability may change across the ABL. To better determine the stability of the whole ABL, the nonlocal definition of static stability can be established based on the kinematic heat flux $\overline{w'\theta_v'}$ at the surface. The ABL is said to be stable, unstable, and neutral if $\overline{w'\theta_v'} < 0$, $\overline{w'\theta_v'} > 0$, and $\overline{w'\theta_v'} = 0$ at the surface, respectively.

A stability parameter that is very useful in the surface layer is the Obukhov length. Starting with Eq. (2.24), nondimensionalize the equation by multiplying it by $-kz/u_*^3$, the buoyancy term becomes $-kzg\overline{w'\theta_v'}/u_*^3\theta_{v0}$. This term is usually assigned a symbol ζ and defined as $\zeta = z/L$, where L is the Obkuhov length. Thus:

$$L = -\frac{u_*^3 \theta_{v0}}{kg\overline{w'\theta_v'}} \tag{2.34}$$

From the above definition, one way to interpret L is that it is proportional to the altitude where buoyant production or destruction of turbulence begins to dominate over mechanical production. In the surface layer, L can be assumed constant, which makes it an important parameter for scaling and similarity arguments. Further discussions are provided in Sect. 2.4.

2.3.3 Dynamic Stability

Dynamic stability suggests whether the flow will become/remain turbulent or laminar. Unlike static stability, dynamic stability is partly influenced by winds. Even in statistically stable ABL, turbulence can be generated dynamically by wind shear. In the statically stable ABL, the buoyant production term of the TKE budget equation (Eq. 2.24) is negative, while the mechanical production term is positive. The ratio between these two terms is a useful nondimensional ratio, called the flux Richardson number:

$$Ri_f = \frac{\frac{g}{\theta_{v0}}\overline{w'\theta_v'}}{\overline{u'w'}\frac{\partial \overline{u}}{\partial z} + \overline{v'w'}\frac{\partial \overline{v}}{\partial z}} \tag{2.35}$$

where the negative sign in the mechanical production term is dropped by convention. If $Ri_f < 1$, the static stability is not strong enough to prevent the generation of mechanical turbulence, and the flow is turbulent (dynamically unstable). If $Ri_f > 1$, the buoyancy consumption dominates over the mechanical production, and the flow becomes laminar (dynamically stable).

However, a problem with Ri_f is that its value can be calculated only for the turbulent flow but not for the intermittently turbulent or laminar flow because it involves the turbulent flux terms ($\overline{u'w'}$, $\overline{v'w'}$, and $\overline{w'\theta_v'}$). To address this, based on the K theory introduced in Sect. 2.1, the flux terms in Eq. (2.35) can be replaced by the corresponding vertical gradients. This gives the gradient Richardson number:

$$Ri = \frac{\frac{g}{\theta_{v0}}\frac{\partial \overline{\theta_v}}{\partial z}}{\left(\frac{\partial \overline{u}}{\partial z}\right)^2 + \left(\frac{\partial \overline{v}}{\partial z}\right)^2} \tag{2.36}$$

Theoretical studies and laboratory experiments indicate that laminar flow becomes turbulent if $Ri < Ri_c$, which is the critical Richardson number usually taken as 0.25. However, for turbulent flow to become laminar, Ri_c may be as high as 1. That is to say, laminar flow needs to decrease below $Ri = 0.25$ to become turbulent, while turbulence can exist up to $Ri = 1$ before the flow becomes laminar.

Meteorological observations are generally made at a series of discrete height intervals. Therefore, the vertical gradients in Eq. (2.36) are usually approximated by finite difference across layers, which gives the bulk Richardson number:

$$Ri_b = \frac{\frac{g}{\theta_{v0}} \Delta\overline{\theta_v} \Delta z}{(\Delta\overline{u})^2 + (\Delta\overline{v})^2} \tag{2.37}$$

Ri_b is a useful indicator of stability, particularly under conditions of low wind shear when Ri becomes unreliable. Note that Ri_f, Ri, Ri_b, and L can also be used to characterize the static stability. The ABL is statically stable if $Ri_f, Ri, Ri_b, \zeta > 0$, unstable if $Ri_f, Ri, Ri_b, \zeta < 0$, and neutral if $Ri_f, Ri, Ri_b, \zeta = 0$.

2.4 Similarity Theory

For a number of ABL situations, it is usually insufficient to derive laws based only on the governing equations. Notwithstanding, observations in the ABL frequently show consistent and repeatable behaviors, which indicates that empirical relationships can be developed for the variables of interest. Similarity theory, which is based on organizing variables into nondimensional groups using Buckingham π dimensional analysis, provides a way for constructing such relationships. Some well-established similarity relationships for the ABL will be introduced in this section. Note that, hereafter, the x-axis is defined in the mean wind direction so that $\overline{v} = \overline{w} = 0$, and only $\overline{u} \neq 0$.

2.4.1 Monin–Obukhov Similarity

Monin–Obukhov similarity theory (MOST) is primarily applied to the surface layer of the atmosphere, therefore this theory is often referred to as surface-layer similarity theory (Monin & Obukhov, 1954; Obukhov, 1971). The surface layer is defined as the portion of the boundary layer where fluxes change by less than 10% with height (see Sect. 1.1). For a first-order approximation, this layer can be treated as a constant flux layer, allowing us to simplify the analysis by using flux measurements taken at a single height, typically at the surface.

The K theory in combination with Prandtl's mixing length hypothesis (Sect. 2.1) gives a logarithmic form of wind profile in the statically neutral surface layer. To extend the analysis to non-neutral surface layer, MOST uses dimensional analysis to construct two nondimensional groups: nondimensional wind shear $\varphi_m = (kz/u_*)\partial U/\partial z$ and nondimensional length scale (also known as Obukhov stability parameter) $\zeta = z/L$. According to MOST, there is a functional dependence of nondimensional wind shear on Obukhov stability parameter, i.e., $\varphi_m = \varphi_m(\zeta)$, which is called the universal function for momentum. $\varphi_m(\zeta)$ can be approximated by the O'KEYPS (named after **O**bukhov, **K**aimal, **E**lliot, **Y**amamoto, **P**anofsky, and **S**ellers) function:

$$[\varphi_m(\zeta)]^4 - \gamma\zeta[\varphi_m(\zeta)]^3 = 1 \tag{2.38}$$

where $\gamma \approx 9$ for $-2 < \zeta < 0$ (Businger et al., 1971). The universal function derived from the Kansas experiment of 1968 by Businger et al. (1971) and later modified by Högström (1988) is presently the most widely used, which is numerically only slightly different from the O'KEYPS function:

$$\text{For } \zeta < 0: \quad \varphi_m(\zeta) = (1 - 19.3\zeta)^{-1/4} \tag{2.39}$$

$$\text{For } \zeta > 0: \quad \varphi_m(\zeta) = 1 + 6\zeta \tag{2.40}$$

Integrating the nondimensional wind shear with height, the logarithmic law for non-neutral surface layer is given as:

$$U = \frac{u_*}{k}\ln\left[\frac{z}{z_0} + \psi_m(\zeta)\right] \tag{2.41}$$

where $\psi_m(\zeta)$ is the integrated form of $\varphi_m(\zeta)$. More details of MOST and its application in modeling the surface layer wind profiles will be given in Sect. 4.2.1.

2.4.2 Rossby Number Similarity

MOST is only applicable to the surface layer. In certain scenarios, such as climate modeling, it is important to connect surface fluxes with external forcings. This requires matching the wind and thermodynamic properties in the upper ABL with those in the surface layer. Consequently, relevant scales include both surface scales and those applicable to the upper ABL (i.e., outer scales).

The geostrophic wind velocity is an appropriate velocity scale in the upper boundary layer. It was proposed that the height scale can be expressed as U_g/f. When this outer scale is paired with the inner scale z_0, it yields the surface Rossby number $Ro = U_g/fz_0$. According to Rossby number similarity, the geostrophic drag coefficient ($C_g = u_*/U_g$, where $U_g = \sqrt{u_g{}^2 + v_g^2}$) and the cross-isobar angle α for the barotropic (i.e., the density depends only on the pressure, and geostrophic wind does not vary with height) and statically neutral ABL is regarded a function of Ro through the following resistance law:

$$A = \ln\left(C_g Ro\right) - \frac{k}{C_g}\cos\alpha \tag{2.42}$$

$$B = \pm \frac{k}{C_g} \sin \alpha \tag{2.43}$$

where A and B are nondimensional coefficients supposed to be universal constants, which were reported to vary between 1.3–1.9 and 4.4–4.9, respectively (Zilitinkevich & Esau, 2002). Minus is applied to the Northern Hemisphere and plus to the Southern Hemisphere in Eq. (2.43).

The resistance law is also applicable to the stratified ABL with the only difference that A and B are no longer constants but nondimensional functions of the Kazanski–Monin stability parameter $\mu = u_*/fL$ in the steady state ABL or h/L (where h is the ABL height) in the evolving ABL. Due to the very wide spread of data in empirical plots of A and B versus μ or h/L, the resistant law was rarely used in practice. Nevertheless, this law recently garnered interest in wind energy studies as they need simple methods to simulate the wind profiles in the whole ABL. Detailed discussions on wind profile models in the ABL will be provided in Chap. 4.

The resistance law is further advanced as Zilitinkevich and Esau (2002) noticed that A and B are not only a function of μ but also of the Brunt–Väisälä frequency $N = \sqrt{(g/\theta_{v0})\partial\theta_v/\partial z}$ in the free atmosphere immediately above the ABL. This involves another stability parameter called the Zilitinkevich number $\mu_N = N/f$. Two types of statically neutral ABL ($\mu = 0$) can then be distinguished: truly neutral ABL, which is developed against a neutrally stratified free flow with $\mu_N = 0$, and conventionally neutral ABL, which is developed against a background stable stratification with $\mu_N > 0$. Similarly, two types of statically stable ABL ($\mu > 0$) can be distinguished: short-lived ABL with $\mu_N = 0$ and long-lived ABL with $\mu_N > 0$.

Truly neutral ABL with $\mu = 0$ and $\mu_N = 0$ is rather rare, which is observed only during short transition periods after sunset on a background of residual layers of convective origin. Therefore, ABLs traditionally treated as neutral are almost always conventionally neutral.

2.4.3 Flux-Variance Similarity

Similar to the TKE equation, the balance equation for the momentum flux can be derived:

$$\frac{\partial \overline{u'w'}}{\partial t} = -\overline{w'^2}\frac{\partial \overline{u}}{\partial z} - \frac{g}{\theta_{v0}}\overline{u'\theta_v'} - \frac{\partial \overline{u'w'^2}}{\partial z} - \frac{1}{\rho}\frac{\partial \overline{w'p'}}{\partial z} - \varepsilon \tag{2.44}$$

From left to right, the terms represent the tendency (term I), the shear production (term II), the buoyancy production or consumption (term III), the turbulent transport (term IV), the pressure transport (term V), and the dissipation (term VI). For the neutral surface layer and steady state condition ($\partial\overline{u'w'}/\partial t = 0$), after estimation of the magnitude of the terms $\overline{w'^2}\partial\overline{u}/\partial z$ and ε, it follows that the nondimensional ratio between the standard deviation of vertical velocity component and the friction velocity σ_w/u_* (where $\sigma_w = \sqrt{\overline{w'^2}}$) is approximately constant. Similarly, σ_u/u_* and σ_v/u_* are approximately constant. These normalized standard deviations are also called integral turbulence characteristics (Foken, 2017). Panofsky and Dutton (1984) recommended $\sigma_u/u_* = 2.45$, $\sigma_v/u_* = 1.9$, and $\sigma_w/u_* = 1.25$, Stull (1988) suggested $\sigma_u/u_* = 2.5$, $\sigma_v/u_* = 2.0$, and $\sigma_w/u_* = 1.25$, while Holmes (2015) used $\sigma_u/u_* = 2.5$, $\sigma_v/u_* = 2.2$, and $\sigma_w/u_* = 1.3 - 1.4$.

A more general form of integral turbulence characteristics for non-neutral conditions is given by:

$$\sigma_i/u_* = c_{1i}(z/L)^{c_{2i}}, \quad i = u, v, w \tag{2.45}$$

where c_{1i} and c_{2i} are constants. Furthermore, based on the Rossby number similarity introduced above, integral turbulence characteristics can also be a function of the Coriolis parameter. Such dependence has been found at least for the neutral condition, and more details are provided in Högström et al. (2002) and Sect. 5.1.1.

2.4.4 Spectral Similarity

Spectral analysis is a powerful tool to help gain a deeper insight into the atmospheric turbulence. In the turbulent part of the spectrum (i.e., periods $<$ 10 min to 2 h), three ranges can be observed. The first range involves the production of turbulent energy by the mean flow, characterized by the integral turbulent length scale (to be introduced in Chap. 5). The typical frequency range is $n \sim 10^{-4}$ Hz (where n is the frequency). This is followed by the inertial sub-range, where turbulence is considered isotropic (meaning turbulent motions lack a preferred direction) and energy decreases in proportion to $n^{-5/3}$. In the third range, with frequencies between 10 and 30 Hz, the kinetic energy of smaller eddies is converted into thermal energy (i.e., energy dissipation). The relevant dissipation length scale here is known as Kolmogorov's microscale $\eta = \left(\nu^3/\varepsilon\right)^{1/4}$ (where $\nu = \mu/\rho$ is the kinematic viscosity), which is about 10^{-3} m (Stull, 1988).

In the inertial subrange, the middle size turbulent eddies are almost not influenced by the generation and dissipation of TKE. These eddies derive energy from larger eddies and, in turn, transfer their energy to smaller eddies. In a steady-state turbulent flow, the rate at which energy cascades down the spectrum must equal the dissipation rate at the smallest eddy sizes. Therefore, there are three variables in this range: S (spectrum), κ (wavenumber, $\kappa = 2\pi/\lambda$, where λ is the wavelength that can be approximated by $\lambda = \overline{u}/n$ based on Taylor's frozen turbulence hypothesis), and ε (dissipation rate). By performing a Buckingham π dimensional analysis, the only nondimensional group is $S^3\kappa^5/\varepsilon^2$, which should be a constant (Stull, 1988). Thus, for the horizontal wind component u spectrum:

$$S_u(\kappa) = \alpha_u \varepsilon^{-2/3} \kappa^{-5/3} \tag{2.46}$$

where α_u is the Kolmogorov constant, which is estimated to be 0.5–0.6.

In the surface layer, the appropriate nondimensional frequency scale would be $f_n = nz/\overline{u}$, the nondimensional spectrum nS_u/u_*^2, and nondimensional dissipation $\varphi_\varepsilon = kz\varepsilon/u_*^3$ (which should be a function only of z/L). Within the inertial subrange, the u spectrum can be expressed as:

$$\frac{nS_u}{u_*^2} = \frac{\alpha_u}{(2\pi k)^{2/3}} \varphi_\varepsilon^{2/3} \left(\frac{nz}{\overline{u}}\right)^{-2/3} \tag{2.47}$$

$\varphi_\varepsilon^{2/3}$ can take the form:

$$\text{For } \zeta < 0: \quad \varphi_\varepsilon^{2/3}(\zeta) = 1 + 0.5|\zeta|^{2/3} \tag{2.48}$$

$$\text{For } \zeta > 0: \quad \varphi_\varepsilon^{2/3}(\zeta) = (1 + 5\zeta)^{2/3} \tag{2.49}$$

More details about the spectral similarity under different stability conditions and in different potions of the ABL are provided in Kaimal and Finnigan (1994).

References

Bolton, J. R., & Hakim, G. J. (2013). *An introduction to dynamic meteorology*. Elsevier Academic Press.

Businger, J. A., Wyngaard, J. C., Izumi, Y., & Bradley, E. F. (1971). Flux-profile relationships in the atmospheric surface layer. *Journal of the Atmospheric Sciences, 28*, 181–189.

Çengel, Y. A., & Cimbala, J. M. (2018). *Fluid mechanics: Fundamentals and applications* (4th ed.). McGraw-Hill Education.

Foken, T. (2017). *Micrometeorology* (2nd ed.). Springer.

Högström, U. (1988). Non-dimensional wind and temperature profiles in the atmospheric surface layer: A re-evaluation. *Boundary-Layer Meteorology, 42*, 55–78. https://doi.org/10.1007/BF00119875

Högström, U., Hunt, J. C. R., & Smedman, A. S. (2002). Theory and measurements for turbulence spectra and variances in the atmospheric neutral surface layer. *Boundary-Layer Meteorology, 103*, 101–124. https://doi.org/10.1023/A:1014579828712

Holmes, J. D. (2015). *Wind loading of structures* (3rd ed.). CRC Press.

Kaimal, J. C., & Finnigan, J. J. (1994). *Atmospheric boundary layer flows: Their structure and measurement*. Oxford University Press.

Monin, A. S., & Obukhov, A. M. (1954). Osnovnye zakonomernosti turbulentnogo peremeshivanija v prizemnom sloe atmosfery (Basic laws of turbulent mixing in the atmosphere near the ground). *Trudy Geofizicheskogo Instituta Akademiya Nauk SSSR, 24*, 163–187.

Obukhov, A. M. (1971). Turbulence in an atmosphere with a non-uniform temperature. *Boundary-Layer Meteorology, 2*, 7–29. https://doi.org/10.1007/BF00718085

Panofsky, H. A., & Dutton, J. A. (1984). *Atmospheric turbulence, models and methods for engineering applications* (1st ed.). Wiley.

Prandtl, L. (1925). Bericht über Untersuchungen zur ausgebildeten Turbulenz. *ZAMM—Journal of Applied Mathematics and Mechanics/Zeitschrift für Angewandte Mathematik und Mechanik, 5*(2), 136–139. https://doi.org/10.1002/zamm.19250050212

Stull, R. B. (1988). *An introduction to boundary layer meteorology* (1st ed.). Kluwer Academic Publishers.

Zilitinkevich, S. S., & Esau, I. N. (2002). On integral measures of the neutral barotropic planetary boundary layer. *Boundary-Layer Meteorology, 104*, 371–379. https://doi.org/10.1023/A:1016540808958

Chapter 3
Wind Observation and Modeling Techniques

This chapter introduces several ways of obtaining wind data. The most traditional way of collecting wind information is using in situ measurements, i.e., instruments deployed directly within the air volume of interest (Sect. 3.1). With the advancement of remote sensing technologies, remote sensing instruments, such as lidars and sodars, have emerged as a powerful tool for wind observations in recent years (Sect. 3.2).

However, observations are usually discrete in space and time. It is impossible to cover every part of the Earth's atmosphere with observations. An alternative approach to reproduce wind fields is to use numerical models, such as numerical weather prediction (NWP) models and computational fluid dynamics (CFD) simulations (Sect. 3.3). They are able to provide gridded, gap-free wind data in three-dimensional space, Albert such simulated data needs careful validation before use.

3.1 In Situ Measurements

In situ measurements are typically conducted using cup anemometers at weather stations and meteorological masts. Sonic anemometers may also be utilized when high-frequency data are required (e.g., for analyzing turbulence intensity, flux, and atmospheric stability).

In situ measurements can usually provide information on mean wind speeds, gust statistics, and turbulence. Cup anemometers generally operate with a temporal resolution up to 1 Hz, whereas sonic anemometers can achieve resolutions up to 100 Hz. Cup anemometers are primarily used to measure hourly or 10-min mean wind speeds and 3-s gust statistics, while sonic anemometers offer detailed wind time series (e.g., highly resolved 10 Hz data) for turbulence analysis.

J. He and Q.-S. Li, *Winds in the Atmospheric Boundary Layer*,
https://doi.org/10.1007/978-981-95-1517-2_3

3.1.1 Cup Anemometer

The cup anemometer is the most commonly used instrument for in situ wind measurement (Fig. 3.1). Typically featuring three or four cups arranged around a vertical axis, it operates by utilizing the difference in flow resistance between the open and closed sides of the cups. This design ensures that the cups rotate consistently in the same direction, irrespective of the wind direction, with their rotational speed being directly proportional to wind speed. As a result, cup anemometers provide average wind speed data. The wind vane, on the other hand, measures the wind direction.

However, cup anemometers have some limitations. They require a minimum wind speed of about 0.3 m/s to start rotating, and they can produce inflated wind speed readings during gusty conditions—an effect known as overspeeding or "u-error". This occurs because the instrument accelerates more quickly than it decelerates. The extent of this overspeeding is influenced by the design of the anemometer, e.g., the distance constant of the cup anemometer.

Cup anemometers are designed to measure horizontal wind, so when wind approaches with an attack angle, the measured wind speed may deviate from the actual speed. This discrepancy is known as the "w-error". This error, which increases under unstable atmospheric conditions, can lead to a horizontal wind speed overestimation of around 10%. Ideally, the measured wind speed should be adjusted based on the cosine of the angle between the wind direction and the horizontal.

Additionally, the discrepancy between scalar and vector mean wind speeds during time averaging may lead to the so-called "DP-error". Variability in wind direction can lead the vector average to be lower than the scalar average, potentially resulting in a discrepancy of up to 10% if the wind direction variance exceeds 30°. Regarding the DP-error, the term "error" may be misleading, as the appropriate average (scalar

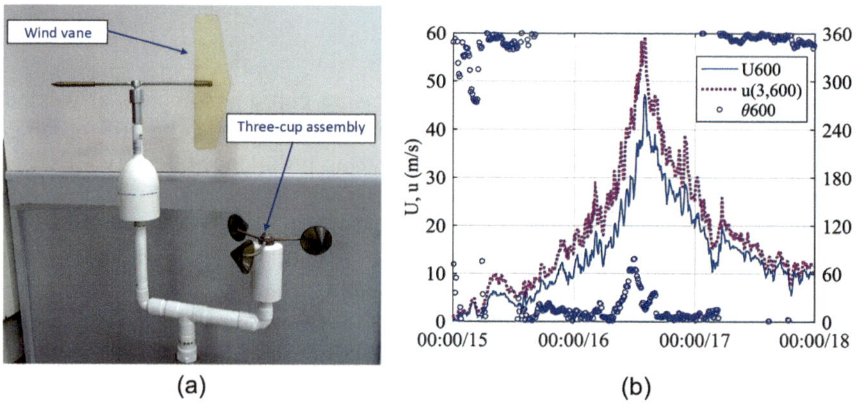

Fig. 3.1 **a** A cup anemometer and wind vane set (image courtesy of the Hong Kong Observatory), and **b** 10-min mean wind speed (blue line), 3-s gust (magenta line), and wind direction (blue dot) recorded by a cup anemometer in a tropical cyclone (He et al., 2020b)

or vector) depends on the specific application for which wind speed is being analyzed (Emeis, 2018).

3.1.2 Sonic Anemometer

A sonic or ultrasonic anemometer (Fig. 3.2) typically features three pairs of ultrasonic transducers that function as both emitters and receivers. Each transducer pair sends sound pulses while measuring the travel time of these pulses. The transducers are usually mounted on thin steel supports, which may cause slight distortions in the airflow. Typically, two pairs are aligned horizontally for horizontal wind measurement, while one pair is positioned vertically for vertical wind measurement. Some designs utilize three inclined paths to reduce flow distortion.

The average sound speed calculated from the signals is proportional to the acoustic temperature. The difference in sound speeds between the two directions for a transducer pair correlates with the wind speed component along that path. While sonic anemometers can accurately measure temperature fluctuations, the mean temperature reading may deviate from the true.

Operating at sound frequencies (around 40–200 kHz), sonic anemometers can capture high-frequency wind fluctuations. They record the travel time of sound between the transducers in both directions, and the difference in these travel times indicates the wind component along the line connecting the transducers. Ultrasonic anemometers can measure at frequencies of up to 100 Hz, making them particularly effective for assessing atmospheric turbulence (Emeis, 2018).

One significant advantage of ultrasonic anemometers is the absence of moving parts, which reduces maintenance requirements and allows for high-resolution data

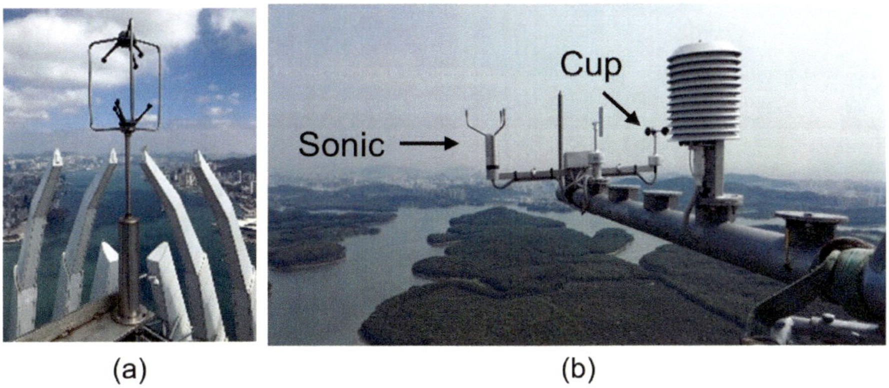

Fig. 3.2 Photos of **a** a sonic anemometer mounted on a tall building (image by the authors), and **b** a cup anemometer and a sonic anemometer mounted on a meteorological mast (image courtesy of the Hong Kong Observatory)

collection across all three wind speed components. Additionally, the covariance between temperature and wind fluctuations provides insights into turbulent heat flux, allowing ultrasonic anemometers to function as both wind and temperature measuring devices, as well as heat flux meters. However, performance may be compromised in rain, as water can alter the acoustic path length.

3.1.3 Other Instruments for in Situ Wind Measurements

Like cup anemometers, propeller (or vane) anemometers (Fig. 3.3) are widely used to measure mean wind speeds. Typically, a propeller anemometer integrates both a propeller and a tail on the same axis, allowing it to measure both wind speed and direction. The speed of the propeller is recorded by a revolution counter, which an electronic chip then converts to wind speed.

Wind speed and direction can also be measured using radiosondes (Figs. 3.4 and 3.5a). A radiosonde is a compact instrument package attached to a helium or hydrogen-filled balloon. As the balloon ascends from the ground, the radiosonde collects data on wind speed and direction (Fig. 3.4b) by tracking its position via GPS (Global Positioning System). The position data obtained during its flight is referred to as a rawinsonde observation. The radiosonde can ascend to over 35 km before the balloon bursts.

Similarly, a dropsonde (Fig. 3.5b) functions like a radiosonde but is specifically designed to be deployed from an aircraft at designated altitudes (e.g., 10 km above mean sea level). It descends to the ground due to gravity, and wind speed and direction are determined by analyzing the GPS coordinates collected during its fall.

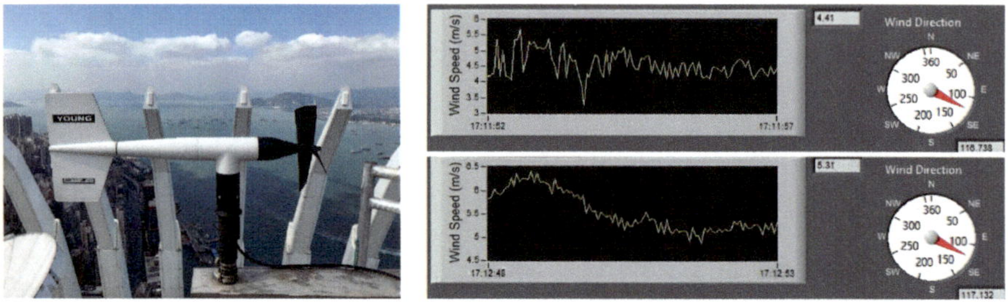

Fig. 3.3 **a** Photo of a propeller anemometer mounted on a tall building, and **b** wind speed recorded by propeller anemometers (image by the authors)

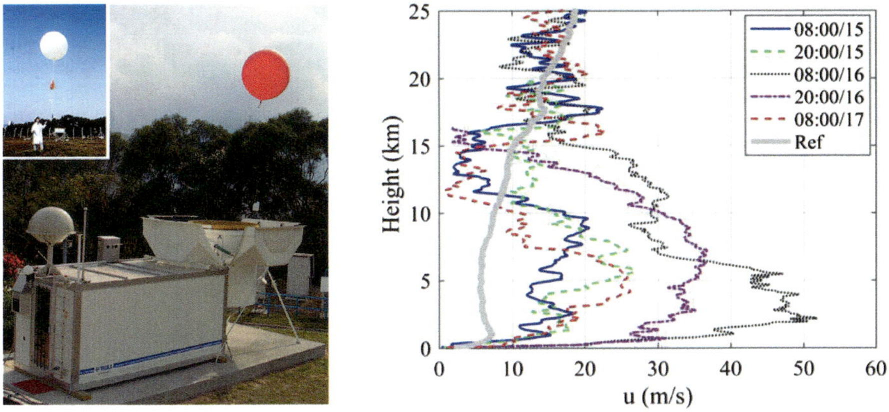

(a) (b)

Fig. 3.4 **a** Launch of a radiosonde (image courtesy of the Hong Kong Observatory), and **b** vertical profiles of wind speeds recorded by radiosondes at various times in a tropical cyclone (He et al., 2020b)

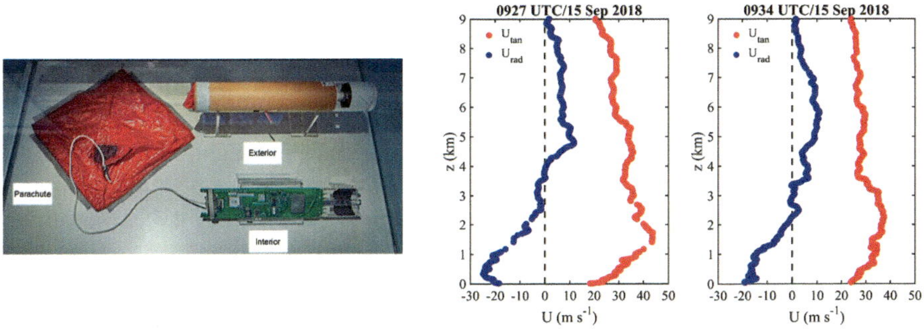

(a) (b)

Fig. 3.5 **a** Components of a dropsonde (image courtesy of the Hong Kong Observatory), and **b** vertical profiles of radial and tangential wind speeds recorded by dropsondes at various times in a tropical cyclone (He et al., 2022b)

3.2 Remote Sensing Observations

3.2.1 Radar Wind Profiler

Radar, which stands for Radio Detection and Ranging, is a type of active remote sensing technology that analyzes backscatters from electromagnetic waves at hydrometeors. In clear air, backscatter is typically unusable unless there are significant moisture gradients, which often occur in clouds. Consequently, radar has been optimized for measuring the position and strength of precipitation, known as weather

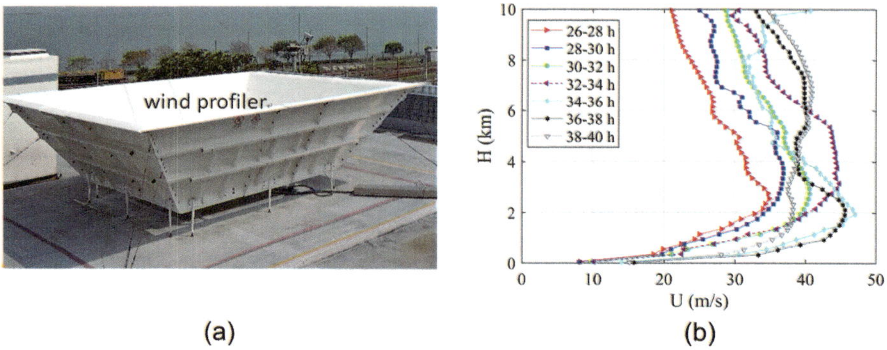

Fig. 3.6 **a** Photo of a wind profiler (image courtesy of the Hong Kong Observatory), and **b** vertical profiles of wind speeds recorded by a wind profiler at various times in a tropical cyclone (He et al., 2020b)

radar. Additionally, by examining the Doppler shift, radar can also be utilized to measure horizontal wind, referred to as Doppler radar.

Wind profilers (Fig. 3.6) are specialized Doppler radar instruments designed to detect vertical wind profiles, operating at frequencies ranging from 50 MHz to 1 GHz. Wind profilers employ Doppler Beam Swinging (DBS) technology to measure three-dimensional wind vectors. This involves sequentially emitting three to five beams in different directions, with one beam directed vertically and others tilted (the azimuthal angles of them typically differ by 90°). Due to the rapid propagation speed of electromagnetic waves (around 3×10^8 m/s), wind profilers achieve a height resolution of about 100 m. The effective range of wind profilers varies with frequency, ranging from 1 to 70 km (Emeis, 2011).

3.2.2 Sodar

The term sodar (Sound Detection and Ranging) is derived in parallel to radar. A sodar emits sound pulses in a nearly vertical direction and examines the backscattered signals that reflect off variations in the atmospheric refractive index. This technology primarily assesses the Doppler shift of the backscattered sound pulses to determine the vertical wind profile, leading to the classification of these devices as Doppler sodars.

Similar to Doppler radar, Doppler sodars (Fig. 3.7) utilize DBS technology to measure wind profiles. They emit three to five beams sequentially in different directions, including a vertical beam and others tilted. The maximum vertical range of the sodar may vary from about 200 m to 1 km, depending on the frequency and emitted power.

Wind speed measurements rely on analyzing the Doppler shift between emitted and received signals. The lowest gate is approximately at a height of 20–30 m, with

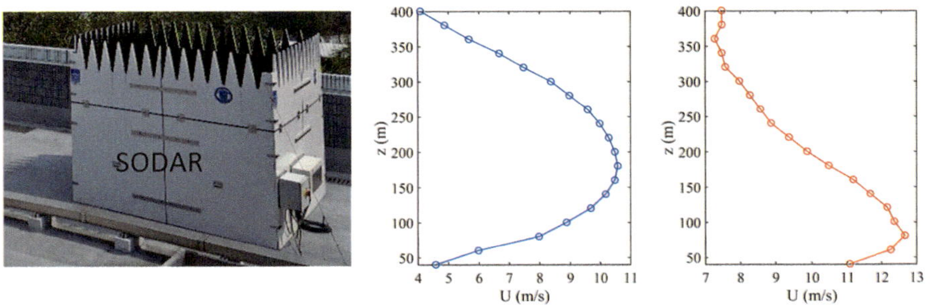

(a) (b)

Fig. 3.7 **a** Photo of a sodar (image courtesy of the Hong Kong Observatory), and **b** vertical profiles of wind speeds recorded by a sodar in two nocturnal low-level jet events (He et al., 2022a)

vertical resolution typically 5–20 m. The backscattered signal represents a specific atmospheric volume. Furthermore, integrating data from multiple beams during a measurement cycle increases the range of the measurement volume. To improve the signal-to-noise ratio, mean winds are calculated over periods of 10–30 min, which makes sodar both a volume- and time-averaging measurement instrument (Emeis, 2011).

3.2.3 Lidar

There are different types of lidar (Light Detection and Ranging) since a light beam can interact with the atmosphere in various ways. This section discusses current optical remote-sensing methods that utilize Doppler shift analysis, particularly focusing on Doppler wind lidars, which are specialized backscatter lidars. Recently, specific Doppler wind lidars designed for measuring wind near the surface (e.g., approximately 40–200 m) have been created for wind energy assessments.

Wind measurement with Doppler lidars (Fig. 3.8) relies on generating two signals with a precise frequency offset. The primary signal comes from a laser known as a master oscillator or pulse laser, while the reference signal is typically emitted from a continuous-wave laser (referred to as a local oscillator). Coherent lidars generally operate in the eye-safe infrared spectrum where aerosol scattering predominates. They determine the Doppler shift by combining the backscattered signal at the receiver with the reference signal at the detector. Due to their sensitivity, coherent lidars are compact yet deliver impressive performance, achieving ranges up to about 20 km, vertical resolutions finer than 100 m, and velocity precisions of tens of cm/s (Emeis, 2011).

Another method for range determination in wind lidars involves beam focusing. This approach assumes that most of the backscattered light originates from the focal range. These lidars (Fig. 3.9) emit a continuous-wave (CW) beam and assess wind

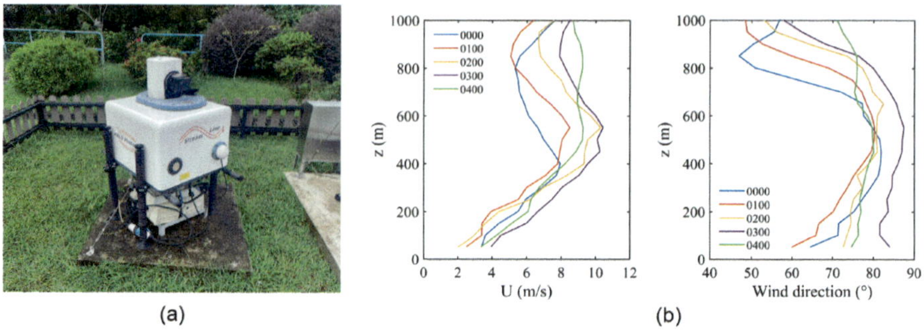

Fig. 3.8 **a** Photo of a coherent lidar (image courtesy of the Hong Kong Observatory), and **b** vertical profiles of wind speed and direction recorded by a coherent lidar at various periods in a nocturnal low-level jet event (He et al., 2024)

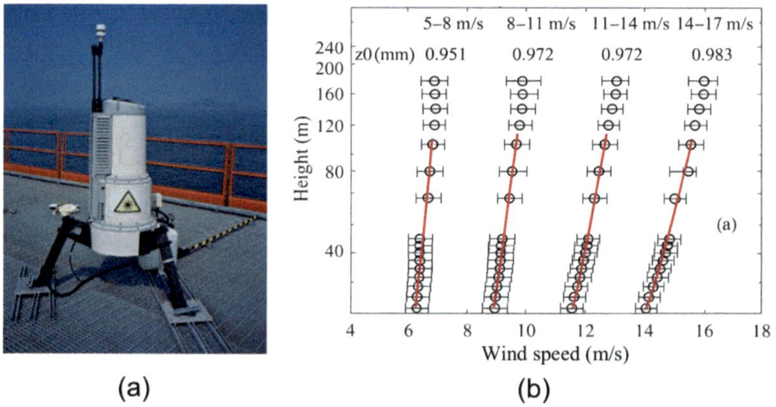

Fig. 3.9 **a** Photo of a continuous-wave lidar, and **b** vertical profiles of mean wind speeds recorded by a continuous-wave lidar (Shu et al., 2016)

speed by focusing, as opposed to the time-of-flight method used in pulsed systems. They can be programmed to measure at various heights, typically between 5 and 150 m, with a height resolution of around 10 m.

3.2.4 Satellite

Wind data can also be obtained from remote sensing instruments installed on weather satellites, including spaceborne microwave radiometers, scatterometers, and synthetic aperture radar (SAR). The algorithms used to retrieve wind data from satellites rely on the observation of wind-generated ocean waves or streaks, which

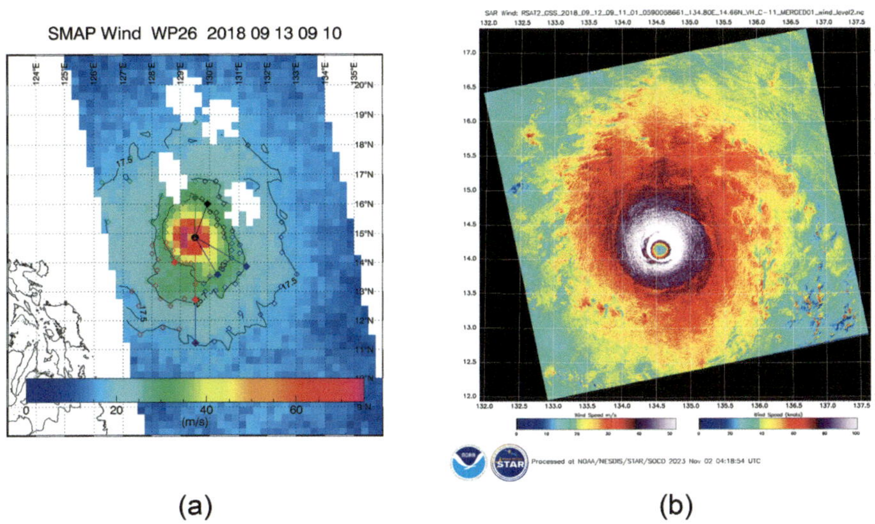

Fig. 3.10 Horizontal spatial distribution of ocean surface wind speeds for a typhoon **a** captured by spaceborne microwave radiometers and **b** captured by SAR (image courtesy of the National Oceanic and Atmospheric Administration)

limits their effectiveness primarily to the ocean surface. Typically, satellite observations offer spatial resolutions ranging from 25 to 50 km (Fig. 3.10a), although SAR technology can achieve much higher resolutions, down to about 100 m (Fig. 3.10b).

Presently available satellite products include ocean surface wind retrievals from Advanced Scatterometer (ASCAT) on MetOp satellites of European Organization for the Exploitation of Meteorological Satellites (EUMETSAT), OceanSat Scatterometer (OSCAT) on Oceansat-2 and ScatSat-1 satellites of Indian Space Research Organization (ISRO), SeaWinds scatterometer on QuikSCAT satellite of National Aeronautics and Space Administration (NASA), WindSat microwave polarimetric radiometer on Coriolis satellite of Naval Research Laboratory (NRL) and Air Force Research Laboratory (AFRL), SARs on Radarsat-2 satellite of Canadian Space Agency (CRA) and Sentinel-1 satellites of European Space Agency (ESA), etc.

3.3 Numerical Models

3.3.1 Numerical Weather Prediction

Numerical weather prediction (NWP) is a method that employs high-speed computers to solve complex mathematical equations representing the governing laws of the atmosphere (e.g., conservation of momentum, mass, and energy, as shown in

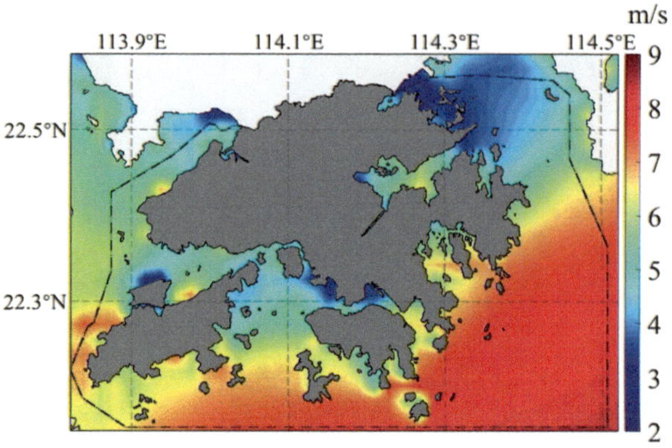

Fig. 3.11 Spatial distribution of mean wind speed at 80-m height over an offshore region from a NWP model (He et al., 2020a)

Sect. 2.1). By establishing initial and boundary conditions, these equations can be solved over time, forming the foundation for weather forecasts.

NWP models, e.g., the Weather Research and Forecasting (WRF) model, simulate atmospheric evolution to provide gridded, gap-free information on variables such as wind (Fig. 3.11), temperature, pressure, and humidity in a three-dimensional (3D) space. In these models, the atmosphere's state at any given moment is represented by meteorological parameters arranged in a 3D grid. A denser grid results in finer model resolution and more detailed forecasts of future atmospheric conditions, but it also significantly increases computational requirements.

However, the current scientific understanding of the atmosphere is not fully sufficient, necessitating approximations and assumptions in the models' equations and methodologies. To this end, parameterizations are employed to estimate the combined effects of physical processes that are too small, brief, complex, or poorly understood to be explicitly included in the governing equations or numerical methods. Each critical physical process that cannot be directly modeled necessitates a parameterization scheme based on sound physical principles or statistical representations. In addition, forecasting remains uncertain due to the chaotic nature of the atmosphere, particularly for rapidly changing and small-scale phenomena like tornadoes, heavy rainfall, and wind shear. These uncertainties typically increase with longer forecast times.

3.3.2 *Reanalysis*

Reanalyses are generated by global NWP models that integrate observations from diverse sources, such as weather stations, buoys, radiosondes, and satellites, by data assimilation. These datasets offer a comprehensive record of global atmospheric conditions. Unlike operational weather prediction, which may vary with changes in methodology, reanalyses are produced using a single version of the modeling system, ensuring consistency.

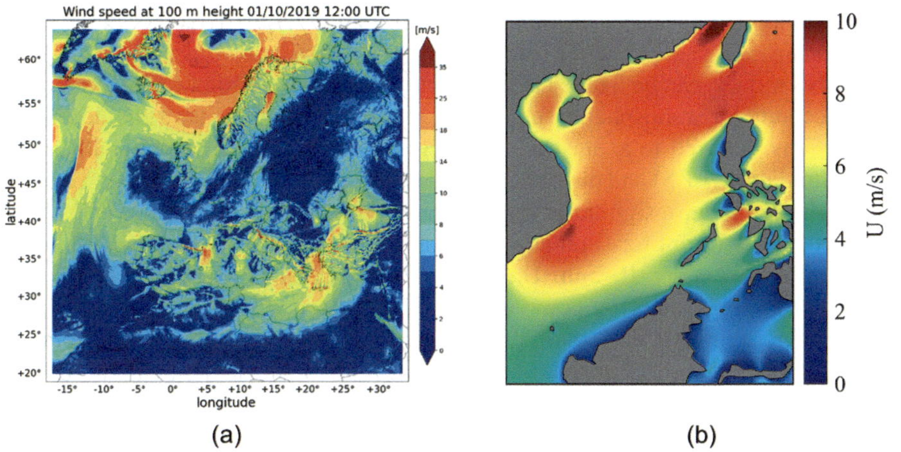

Fig. 3.12 Spatial distribution of **a** hourly wind speed at 100-m height over Europe and **b** 20-year mean wind speed at 100-m height over South China Sea from ERA5 (He et al., 2023)

Typically, reanalysis data is coarsely resolved (with horizontal spatial resolution > 30 km), making it unsuitable for capturing small-scale wind fluctuations at specific locations. As a result, reanalysis data is best used for larger spatial and temporal scales. For small-scale simulations, higher-resolution NWP models can be employed, utilizing the large-scale reanalysis data as input. Since reanalyses provide gridded data, discrepancies can arise when comparing them to station observations. These differences are influenced not only by scale but also by the specific numerical weather model and the volume of observations incorporated into the assimilation process.

Publicly available reanalysis datasets include ERA-Interim, ERA5 (Fig. 3.12), and ERA-20C from the European Centre for Medium-Range Weather Forecasts (ECMWF), as well as NCEP-NCAR, NCEP-GFS, NCEP-FNL, and NCEP-CFSv2 from the National Center for Environmental Prediction (NCEP). Other notable datasets include MERRA-2 from National Aeronautics and Space Administration (NASA) and JRA-25 and JRA-55 from the Japan Meteorological Agency (JMA).

3.3.3 Computational Fluid Dynamics

NWP models typically operate in an unsteady mode to make forecasts or perform scenario analyses. In contrast, computational fluid dynamics (CFD) models often focus on a simpler equilibrium between pressure gradient forces and frictional forces, running in a stationary mode under specified external conditions. These stationary CFD models are known as Reynolds Averaged Navier–Stokes (RANS) models, as they rely on numerical solutions of the Reynolds-averaged Navier–Stokes equations (see Sect. 2.1), with turbulence usually parameterized. A common approach is to use a two-equation model based on the *k* (turbulent kinetic energy)–*e* (energy dissipation) equations.

RANS models can be likened to wind tunnels. More sophisticated RANS models may account for atmospheric stability in the ABL or allow for time-dependent simulations, referred to as unsteady RANS (URANS) models. Additionally, large eddy simulation (LES) and direct numerical simulation (DNS) have been developed, although they require significant computational resources. LES models (Fig. 3.13) are particularly useful for simulating phenomena at spatial scales below 100 m and temporal scales under 10 min. Unlike traditional RANS models, LES explicitly models the energy-carrying components of turbulence while parameterizing only the smaller subgrid-scale turbulence. Although they are well-suited for capturing unsteady processes, LES models are resource-intensive and are typically limited to simulations lasting from a few minutes to several hours.

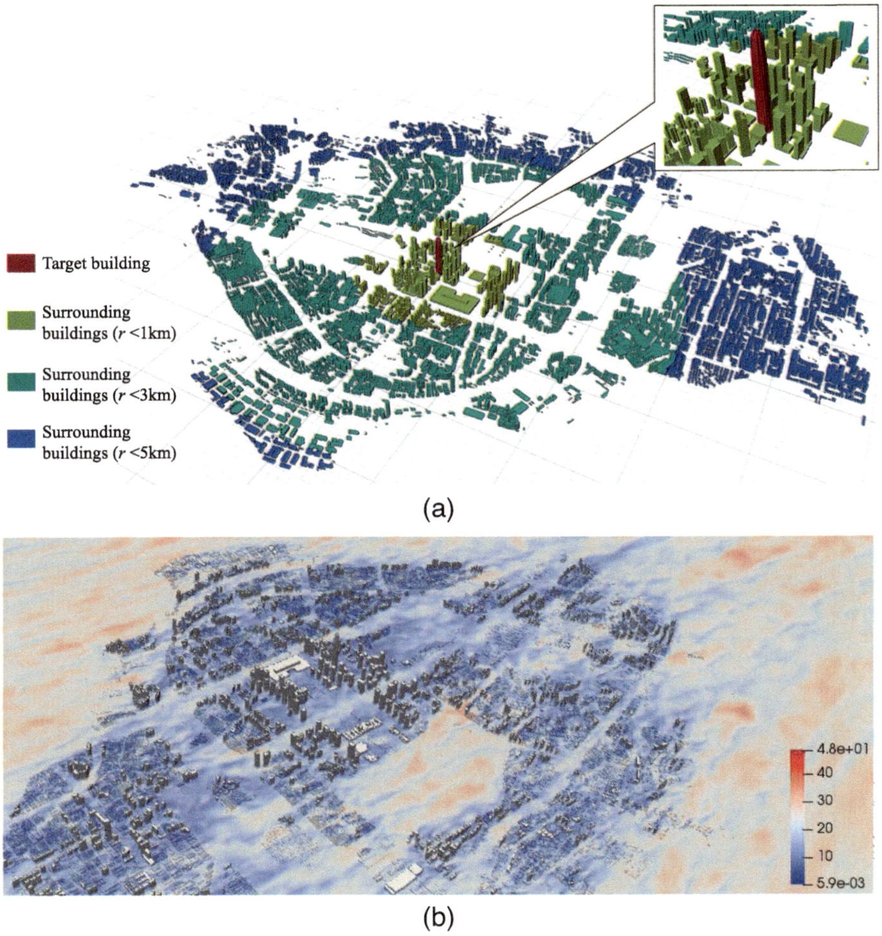

Fig. 3.13 LES simulation of wind speed over a high-density urban area, **a** building models, and **b** spatial distribution of wind speed (Lu et al., 2024)

References

Emeis, S. (2011). *Surface-based remote sensing of the atmospheric boundary layer*. Springer.

Emeis, S. (2018). *Wind energy meteorology—Atmospheric physics for wind power generation* (2nd ed.). Springer.

He, J. Y., Chan, P. W., Li, Q. S., & Lee, C. W. (2020a). Spatiotemporal analysis of offshore wind field characteristics and energy potential in Hong Kong. *Energy, 201*, 117622. https://doi.org/10.1016/j.energy.2020.117622

He, J. Y., Chan, P. W., Li, Q. S., & Lee, C. W. (2022a). Characterizing coastal wind energy resources based on sodar and microwave radiometer observations. *Renewable and Sustainable Energy Reviews, 163*, 112498. https://doi.org/10.1016/j.rser.2022.112498

He, J. Y., Chan, P. W., Li, Q. S., & Tong, H. W. (2023). Mapping future offshore wind resources in the South China Sea under climate change by regional climate modeling. *Renewable and Sustainable Energy Reviews, 188*, 113865. https://doi.org/10.1016/j.rser.2023.113865

He, J. Y., Chan, P. W., Liu, Z. F., Fung, J. C. H., & Li, Q. S. (2024). Observation of wind and thermodynamic structure within an urban boundary layer. *Physics of Fluids, 36*(8), 086603. https://doi.org/10.1063/5.0214961

He, J. Y., Hon, K. K., Li, Q. S., & Chan, P. W. (2022b). Wind profile analysis for selected tropical cyclones over the South China Sea based on dropsonde measurements. *Atmósfera, 35*, 111–126. https://doi.org/10.20937/atm.52900

He, Y. C., He, J. Y., Chen, W. C., Chan, P. W., Fu, J. Y., & Li, Q. S. (2020b). Insights from Super Typhoon Mangkhut (1822) for wind engineering practices. *Journal of Wind Engineering & Industrial Aerodynamics, 203*, 104238. https://doi.org/10.1016/j.jweia.2020.104238

Lu, B., Li, Q. S., Wang, X. C., Han, X. L., & He, J. Y. (2024). Large eddy simulation of wind pressures on a 600-m-high skyscraper and comparison with field measurements during Super Typhoon Mangkhut. *Journal of Building Engineering, 92*, 109750. https://doi.org/10.1016/j.jobe.2024.109750

Shu, Z. R., Li, Q. S., He, Y. C., & Chan, P. W. (2016). Observations of offshore wind characteristics by Doppler-LiDAR for wind energy applications. *Applied Energy, 169*, 150–163. https://doi.org/10.1016/j.apenergy.2016.01.135

Chapter 4
Mean Wind Profiles Over Homogenous Terrain

This chapter discusses current progress in observing and modeling the vertical profile of mean wind speed, or mean wind profile, in the atmospheric boundary layer over homogenous terrain. For design of civil structures against wind actions, neutral atmospheric stability conditions (adiabatic atmosphere; Sect. 4.1) are generally assumed due to the high wind shear and strong mechanical turbulence in such conditions (see Sect. 2.3 for the concept of atmospheric stability). However, for other applications such as pollutant dispersion modeling and wind energy resource assessment, it is essential to consider both neutral and non-neutral conditions (both adiabatic and diabatic atmosphere; Sect. 4.2). In addition, the best practice of wind-resistant structural design needs to account for the specific shapes of wind profiles in severe windstorms, such as tropical cyclones and thunderstorms (Sect. 4.3).

4.1 Wind Profiles in Neutral Conditions

Wind pressure acting on structures is directly proportional to the square of wind speed. Therefore, an accurate description of the vertical profile of horizontal mean wind speeds up to altitudes of several hundred meters in neutral atmospheric stability conditions is of central importance for design of buildings and wind turbines against wind actions.

4.1.1 Power Law

One of the most popular wind profile models used in structural design and wind energy applications is the power law. The power law is fundamentally empirical and lacks a theoretical foundation, yet it offers the advantage of straightforward integration over height. This feature is particularly useful for calculating bending

J. He and Q.-S. Li, *Winds in the Atmospheric Boundary Layer*,
https://doi.org/10.1007/978-981-95-1517-2_4

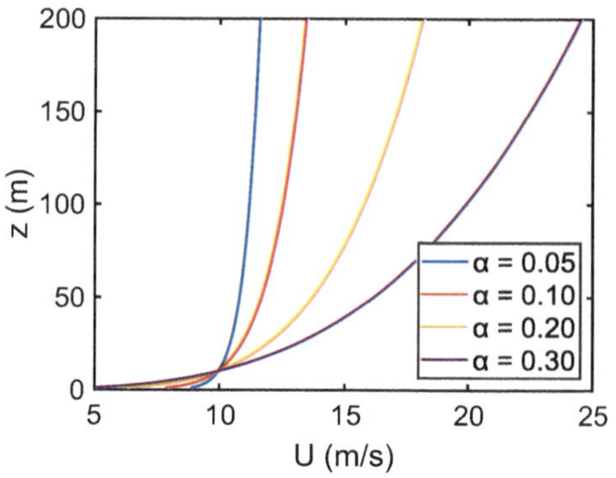

Fig. 4.1 Power law wind profiles with different power law exponents

moments at the base of tall structures. The concept was initially introduced nearly a century ago for open terrains and urban areas. Its simplicity has led to its widespread adoption in structural design codes and standards for representing the neutral ABL wind profile, which remains prevalent in the United States, Canada, Japan, and China. The power law is expressed as:

$$U = U_1\left(\frac{z}{z_1}\right)^{\alpha} \tag{4.1}$$

where U is the mean wind speed (usually 10-min mean or hourly mean) at height z, U_1 is the mean wind speed at reference height z_1, e.g., 10 m height above ground level, and α is the power law exponent (also known as the wind shear coefficient and the Hellmann exponent). Commonly used values of power law exponents are listed in Sect. 4.1.4. Figure 4.1 shows power law wind profiles with different power law exponents, assuming a wind speed of 10 m/s at a height of 10 m.

According to design codes, the power law is considered valid up to a height referred to as the "gradient" height, which is believed to represent the ABL height. When applied to 3-s peak gust wind profiles, the power law maintains the same formulation as Eq. (4.1), while the power law exponent may vary. Since the power law is not grounded on any ABL theories or assumptions, it is not restricted to the neutral ABL and can be utilized in both neutral and non-neutral atmospheric stability conditions (see Sect. 4.2).

4.1.2 Logarithmic Law

The logarithmic law is much more widely accepted by the meteorological community due to its stronger theoretical foundation. Originally derived by Prandtl for the turbulent boundary layer over a flat plate, it has been shown to hold true, in its unaltered form, in strong wind scenarios within the ABL close to the surface.

There are various methods to derive the logarithmic law. One approach utilizes the mixing length hypothesis (see Sect. 2.2.1), while another, which is introduced in the following, employs dimensional analysis. To estimate the mean wind speed as a function of height above the ground, we consider two relevant variables: surface stress (represented by the friction velocity u_*) and surface roughness (represented by the roughness length z_0). By applying dimensional analysis, we can identify two dimensionless groups: U/u_* and z/z_0. Based on experimental data, we anticipate a logarithmic relationship between these two dimensionless groups (Stull, 1988):

$$U = \frac{u_*}{k}\ln\left(\frac{z}{z_0}\right) \tag{4.2}$$

where k is the von Kármán constant mostly taken as 0.4, and z_0 is the roughness length. The logarithmic law is established for the surface layer only (which is the bottom 10% of the ABL; see Sect. 1.1), where the friction velocity u_* generally varies by less than 10%. It is not applicable to the upper part of the ABL (i.e., Ekman layer).

Over rough terrain such as urban environments or forests, individual roughness elements (such as buildings and trees) are densely clustered together. In this context, the tops of these elements function as a displaced surface. Consequently, the reference level for the wind profile, as described by the logarithmic law, is not at the ground level but rather at a height d above the surface within the canopy:

$$U = \frac{u_*}{k}\ln\left(\frac{z - z_d}{z_0}\right) \tag{4.3}$$

where z_d is referred to as zero-plane displacement or displacement height. Consequently, all equations for the mean wind profile and turbulence characteristics must be modified by replacing z with $z - z_d$. Figure 4.2 shows logarithmic law wind profiles with different roughness lengths and displacement heights, assuming a wind speed of 10 m/s at $z - z_d = 10$ m.

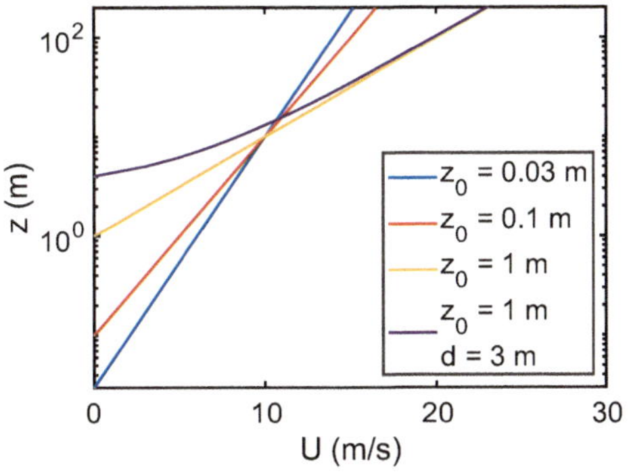

Fig. 4.2 Logarithmic law wind profiles with different roughness lengths and displacement heights (height axis in logarithmic scale)

While the logarithmic law is theoretically robust, particularly for fully developed wind flow over uniform terrain, such ideal conditions are seldom encountered in real-world conditions. Additionally, the logarithmic law has certain mathematical limitations that can lead to complications. First, since logarithms of negative numbers are undefined, the law cannot be applied for heights below the zero-plane displacement. Second, integrating the logarithmic law is more complex. To mitigate these issues, wind engineers often opt for the simpler power law.

4.1.3 Wind Profiles Above the Surface Layer

The logarithmic law is restricted to the surface layer. However, modern tall buildings and wind turbines are often situated partially within the Ekman layer. This layer, situated above the surface layer, remains part of the ABL due to the continued influence of surface friction. Consequently, in addition to the vertical profiles in the surface layer, assessing wind loads requires consideration of the more complex wind dynamics present in the Ekman layer.

When ascending from the surface layer into the Ekman layer, the balance of forces shifts. In this layer, not only are the pressure gradient force and surface friction significant, but the Coriolis force, resulting from the Earth's rotation, also becomes crucial. These three forces must achieve equilibrium in a stationary Ekman layer. As a result, the Ekman layer is characterized by a distinct change in wind direction with height.

In the Ekman layer, it is assumed that the height-dependent growth of the exchange coefficient K_m ceases at the top of the surface layer, becoming constant throughout the remainder of the boundary layer (see Sect. 2.2.2). This presents challenges when trying to unify the wind profile description across both layers. Due to the constant K_m assumption in the Ekman layer, certain relations cannot be extended downward into the surface layer, and vice versa, because of the mixing length's linear growth in the surface layer (see Sect. 2.2.1).

To address these challenges, two approaches have been proposed. The first involves creating a smooth transition in wind speed and shear between the surface layer and Ekman layer profiles (i.e., profile matching). By this means, Deaves and Harris (1978) gave the following engineering model of wind profiles in the ABL:

$$U = \frac{u_*}{k}\left[\ln\left(\frac{z}{z_0}\right) + 5.75\frac{z}{h} - 1.88\left(\frac{z}{h}\right)^2 - 1.33\left(\frac{z}{h}\right)^3 + 0.25\left(\frac{z}{h}\right)^4\right] \tag{4.4}$$

where the ABL height is given as $h = u_*/6f$.

The second approach focuses on modifying the mixing length to extend the surface layer wind profile into higher layers. Blackadar (1962) developed the following model based on a length scale that increases linearly with height near the surface and approaches a constant value at upper levels:

$$U = \frac{u_*}{k}\left[\ln\left(\frac{z}{z_0}\right) + \frac{kz}{\eta} - \frac{z}{h}\left(\frac{kz}{2\eta} + 1\right)\right] \tag{4.5}$$

$$\eta = \frac{k_b u_*}{f} \tag{4.6}$$

where $k_b = 0.0063$.

More recently, Gryning et al. (2007) reformulated the height-dependent mixing length to restrict its growth with increasing height. They parameterized the length scales in the surface layer, middle boundary layer, and upper boundary layer so as to extend the applicability of the logarithmic law beyond the surface layer:

$$U = \frac{u_*}{k}\left[\ln\left(\frac{z}{z_0}\right) + \frac{z}{L_{MBL,N}} - \frac{z}{h}\left(\frac{z}{2L_{MBL,N}}\right)\right] \tag{4.7}$$

$$L_{MBL,N} = \frac{h}{2\left[\left[(\ln(u_*/fz_0) - A)^2 + B^2\right]^{1/2} - \ln(h/z_0)\right]} \tag{4.8}$$

where A and B are constants from the resistance law, which were reported to vary between 1.3–1.9 and 4.4–4.9 for neutral ABL, respectively (Zilitinkevich & Esau, 2002).

Examples of wind profiles described by these models are given for a site with the following parameters: $z_0 = 0.03$ m (open terrain), $f = 7.3 \times 10^{-5}\,\text{s}^{-1}$ (latitude of 30°). For a storm producing $u_* = 1$ m/s, the ABL depth is estimated to be $h = u_*/6f = 2.3$ km. The wind profiles described by the aforementioned models are shown in Fig. 4.3, where Log: Logarithmic; B62: Blackadar (1962); D78: Deaves and Harris (1978); G07: Gryning et al. (2007).

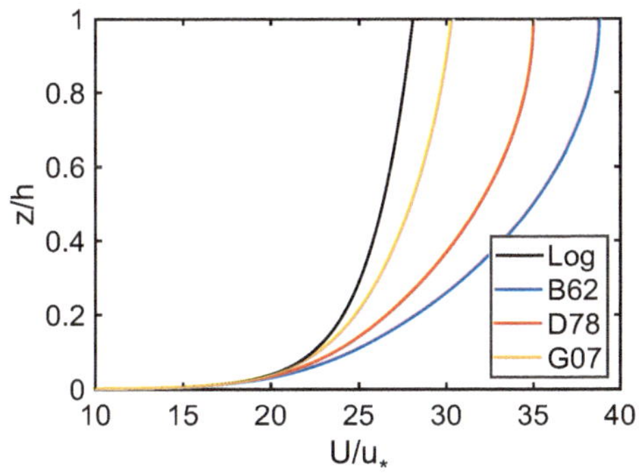

Fig. 4.3 Wind profiles over open terrain described by the logarithmic law, Deaves and Harris model, Blackadar model, and Gryning model

4.1.4 Surface Roughness

The roughness length introduced in previous sections is the height at which wind speed reaches zero. While the roughness length does not equate to the height of individual surface roughness elements, there is a direct relationship between them. Once established for a specific surface, the roughness length remains constant regardless of variations in wind speed, stability, or stress. However, it can change if the surface roughness elements themselves are altered, such as through changes in vegetation height and coverage, the construction of fences or buildings, deforestation, or logging activities.

Typical values for the roughness length are illustrated in Table 4.1 (Wieringa, 1992). Generally, taller roughness elements correspond to larger aerodynamic roughness lengths, although in all cases, the aerodynamic roughness length is smaller than the actual height of the roughness elements.

Over oceans, the roughness length is associated with the ocean wave height, which is in turn a function of surface stress or wind speed. The ocean surface roughness length can be parameterized by the Charnock relation (Charnock, 1955):

$$z_0 = \alpha_c \frac{u_*^2}{g} \tag{4.9}$$

where α_c is the Charnock coefficient of a range 0.015–0.035, which is higher for shallow coastal waters while lower for deep open oceans.

Table 4.1 Roughness length and zero-plane displacement for typical surfaces from Wieringa (1992)

Surface type	Roughness length (m)
Sea, loose sand and snow	0.0002 (wind speed-dependent)
Concrete, flat desert, tidal flat	0.0002–0.0005
Flat snow field	0.0001–0.0007
Rough ice field	0.001–0.012
Fallow ground	0.001–0.004
Short grass and moss	0.008–0.03
Long grass and heather	0.02–0.06
Low mature agricultural crops	0.04–0.09
High mature crops ("grain")	0.12–0.18
Continuous bushland	0.35–0.45
Mature pine forest	0.8–1.6
Dense low buildings ("suburb")	0.4–0.7
Regularly-built large town	0.7–1.5
Tropical forest	1.7–2.3

However, the Charnock relation tends to underestimate sea surface roughness length at low wind speeds, as the capillary waves in these conditions are particularly rough. To improve accuracy at low wind speeds, the following relationship proposed by Zilitinkevich (1969) may be used:

$$z_0 = c_1 \frac{\nu}{u_*} + \frac{u_*^2}{c_2 g} \tag{4.10}$$

where ν is the kinematic viscosity of air, and the coefficients $c_1 = 0.1$ and $c_2 = 20.8$.

For rough surfaces like urban areas, forests, and dense vegetation, the canopy can be treated as a porous medium, necessitating consideration of zero-plane displacement. A simple estimation for zero-plane displacement is to take it as approximately 2/3 of the canopy height:

$$z_d = \frac{2}{3} h_c \tag{4.11}$$

where h_c is the canopy height.

While the logarithmic law is more theoretically correct, the power law is more convenient for practical use. By matching the logarithmic law with the power law, Panofsky and Dutton (1984) showed that the power law exponent can be estimated by the roughness length, as follows:

$$\alpha = \frac{1}{\ln(z_{ref}/z_0)} \tag{4.12}$$

where z_{ref} is the matching height between the logarithmic law and the power law. z_{ref} may be taken as $z_{ref} = \sqrt{z_1 z_2}$, were z_1 and z_2 are the lowest and highest levels of observations, respectively. Figure 4.4 shows an example of such matching over open terrain, with $z_0 = 0.03$ m, $z_1 = 10$ m, and $z_2 = 100$ m.

In wind or structural design codes and standards, the power law is the most popular one, followed by the logarithmic law. The power law exponents and roughness lengths

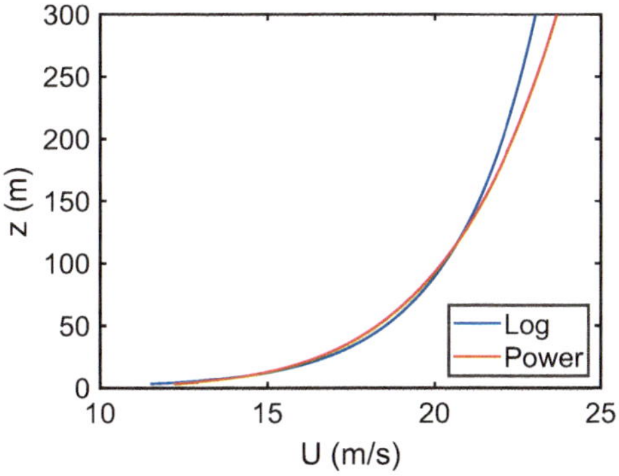

Fig. 4.4 Matching of power law wind profiles with logarithmic law wind profiles

stipulated in major wind codes and standards for different terrain (or exposure) categories are summarized in Table 4.2, where "ASCE" represents Minimum Design Loads for Buildings and Other Structures (ASCE, 2022), "AIJ" represents Recommendations for Loads on Buildings (AIJ, 2015), "AS/NZS" represents Structural Design Actions (AS/NZS, 2011), "CNS" represents Chinese Load Code for the Design of Building Structures (CNS, 2012), "NBCC" represents National Building Code of Canada (NRC, 2015), "EU" represents Eurocode 1: Actions on structures (CEN, 2005), and "ISO" represents Wind Actions on Structures (ISO, 2009). The exposure categories are classified as I: city center with heavy concentration of tall buildings; II: general urban area; III: suburban area; IV: open terrain area where basic wind velocity is defined; V: open lake or land with minimal obstruction; and VI: sea or coastal area (Kwon & Kareem, 2013).

Figure 4.5 shows the wind profiles for different terrain categories stipulated in the Chinese Load Code for the Design of Building Structures, assuming a gradient wind speed of 50 m/s. The power law exponents are 0.12, 0.15, 0.22, and 0.3 for sea, country, city, and city center, respectively, and the gradient heights are 300, 350, 450, and 550 m, respectively.

Given the governing role of surface roughness in the wind speed and turbulence profiles, an accurate estimation of the roughness length is of great significance. Two

Table 4.2 Power law exponents and roughness lengths in major wind codes and standards

Exposure category	ASCE α	AIJ α	CNS α	NBCC α	AS/NZS z_0 (m)	EU z_0 (m)	ISO z_0 (m)
I	–	0.35	–	–	2	–	3
II	–	0.27	0.3	0.36	–	1	–
III	0.25	0.2	0.22	0.25	0.2	0.3	0.3
IV	0.15	0.15	0.15	0.14	0.02	0.05	0.03
V	0.1	0.1	0.12	–	0.002	0.01	0.003
VI	–	–	–	–	–	0.003	–

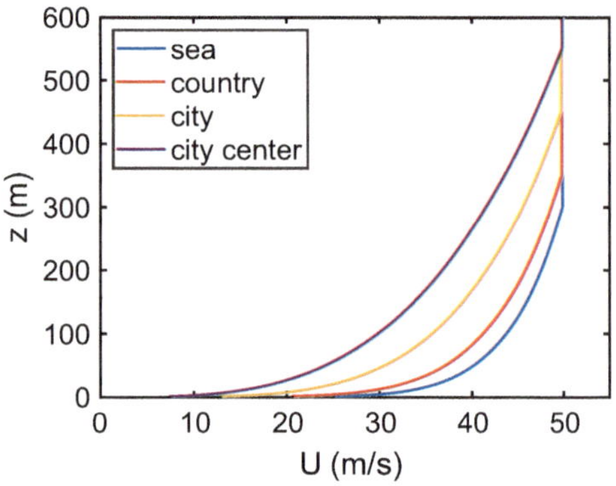

Fig. 4.5 Wind profiles for different terrain categories stipulated in the Chinese Load Code for the Design of Building Structures

major categories of estimation methods include morphometric methods and anemometric methods. Morphometric methods mostly rely on information on roughness elements such as satellite imagery. One of the simplest morphometric methods was proposed by Lettau (1969):

$$z_0 = 0.5h^* \frac{s_S}{s_L} \tag{4.13}$$

where h^* is the mean height of the roughness elements, s_s is the mean cross-sectional area of a single element facing the wind, and s_s is the lot size per element. This approach is valid under conditions where the roughness elements are evenly spaced, not overly dense, and have similar height and shape.

Anemometric methods are based on wind observations. One of the most widely used anemometric methods is the profile method, which is applicable to locations where multi-level wind records in the neutral surface layer (usually below a height of 100 m) are available. The roughness length can be determined by firstly plotting the natural logarithm of heights against wind speeds, then fitting a straight line to the profile and extrapolating downward to the surface (Barnéoud & Ek, 2020). An example of such fitting to determine the speed-dependent roughness lengths of a sea surface is shown in Fig. 4.6.

Another anemometric method is the flux or eddy covariance method, which can be applied if measurements of three orthogonal fluctuating wind components are available (Grimmond et al., 1998). Assuming neutral stability condition, the inversion of the logarithmic law yields:

$$z_0 = z \times \exp\left(-\frac{kU}{u_*}\right) \tag{4.14}$$

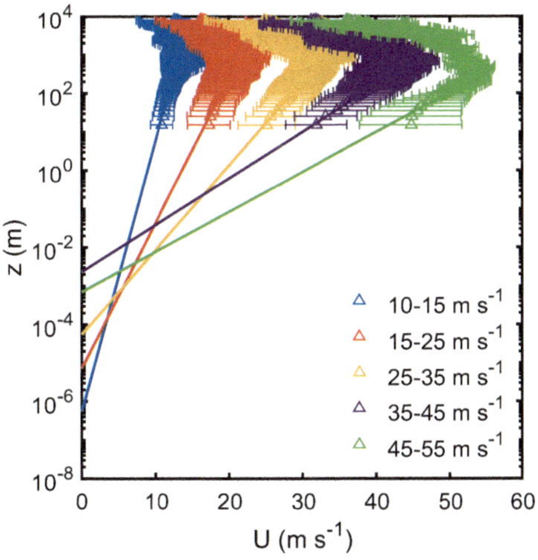

Fig. 4.6 Fitting of wind profiles to determine roughness lengths of a sea surface

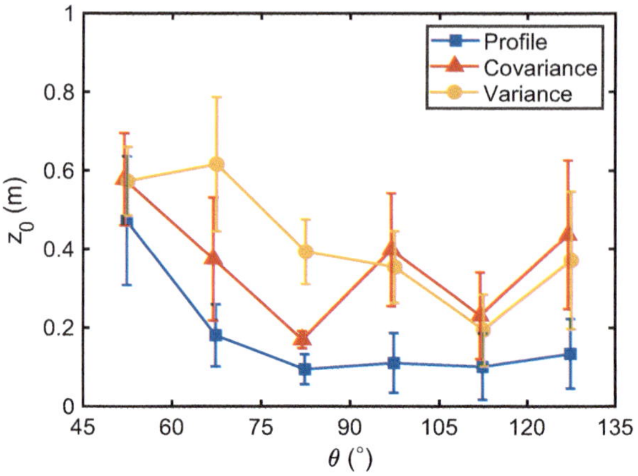

Fig. 4.7 Estimation of roughness lengths by the profile method, eddy covariance method, and variance method, for different wind directions around a tall meteorological tower (He et al., 2020b)

where u_* is evaluated directly by $u_* = \left(\overline{u'w'}^2 + \overline{v'w'}^2\right)^{1/4}$ using the measurements of three orthogonal fluctuating wind components (Foken, 2017).

When multi-level wind speeds and three orthogonal fluctuating wind components are not available, the variance method could be utilized as it simply requires records from fast-response anemometry at a single height level (Giammanco et al., 2016). z_0 can be estimated by:

$$z_0 = z \times \exp\left(-\frac{kU}{\sigma_i}\frac{\sigma_i}{u_*}\right) = z \times \exp\left(-\frac{k}{I_i}\frac{\sigma_i}{u_*}\right), \quad i = u, v, w \tag{4.15}$$

where $I_i = \sigma_i/U$, $i = u, v, w$ are the turbulence intensity (to be introduced in Sect. 2.6.2). $\sigma_u/u_* = 2.45$, $\sigma_v/u_* = 1.9$, and $\sigma_w/u_* = 1.25$ can be used in the surface layer (Foken, 2017). Results from different methods are usually different because of the measurement errors and the different assumptions made. Figure 4.7 shows the estimation of roughness lengths by the profile method, eddy covariance method, and variance method, for different wind directions around a tall meteorological tower over a rural area.

4.2 Wind Profiles in Non-neutral Conditions

4.2.1 Wind Profiles in Unstable and Stable Conditions

Wind profiles in neutral atmospheric stability conditions have been introduced in Sect. 4.1, which are useful for wind-resistant structural design. Nevertheless, neutral ABL is only found in very strong winds or during short transition periods after sunset, while stable or unstable conditions occur more frequently.

Figure 4.8 portrays a typical evolution of the ABL during a diurnal cycle. After sunrise, heat flux from the surface warms the overlying air, leading to a statically unstable ABL. This fosters the rapid formation of a convective boundary layer (or mixed layer), which is highly turbulent and well-mixed, capped by an entrainment zone above. Shortly before sunset, a stable nocturnal boundary layer forms close to the ground, characterized by a surface inversion. Above this layer, remnants of the daytime mixed layer persist, exhibiting weaker turbulence and referred to as the residual layer, which is capped by a free (capping) inversion (Foken, 2017).

The logarithmic law presented in Sect. 4.1.2 is valid specifically within the neutral surface layer, where the effects of thermal stratification can be ignored. Monin and Obukhov (1954) applied dimensional analysis to broaden these equations for non-neutral conditions. In their analysis, the dependent parameters in the surface layer include height z, friction velocity u_*, kinematic heat flux $\overline{w'\theta_v'}$, and the buoyancy parameter g/θ_v. The independent dimensions are length, time, and temperature. Based on this analysis, a key dimensionless parameter that describes the dynamics of the surface layer is the Obukhov stability parameter $\zeta = z/L$. The characteristic length scale, known as the Obukhov length L, relates dynamic, thermal, and buoyancy processes and is proportional to the height of the surface layer.

As given in Sect. 2.3.3, the Obukhov length can be derived from the turbulent kinetic energy (TKE) equation. A physical interpretation of this length is that the absolute value of the Obukhov length corresponds to the height of an air column where the production (for $L < 0$) or loss (for $L > 0$) of TKE due to buoyancy forces equals the dynamic production of TKE per unit volume at any height, multiplied by the height z.

For considerations of buoyancy, water vapor content is significant as it affects air density. Therefore, for accurate calculations, the Obukhov length is ideally calculated

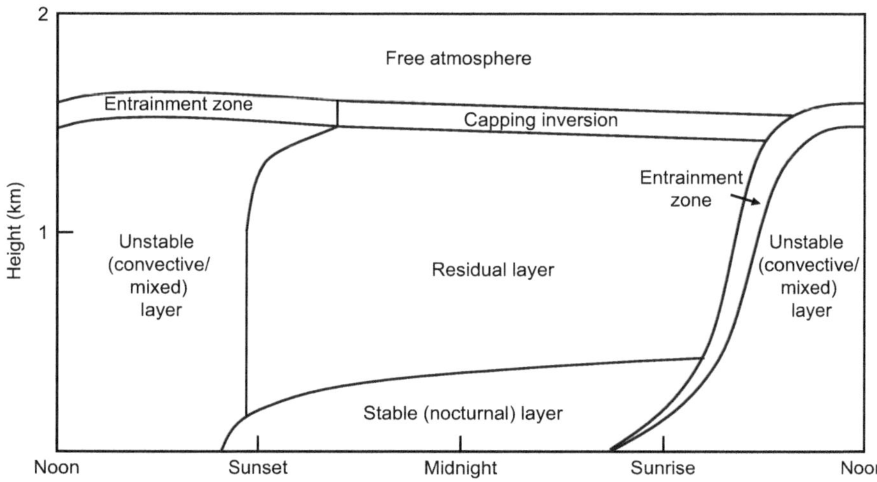

Fig. 4.8 Typical diurnal variation of the ABL

using the virtual potential temperature θ_v rather than the potential temperature θ. However, θ is more commonly employed than θ_v because the universal functions relevant to the Obukhov length were primarily derived from observations in the lower surface layer, often within dry regions. Observations in other areas have not revealed significant discrepancies within measurement accuracy.

As introduced in Sect. 2.4.1, the Monin–Obukhov similarity theory is applied to the wind profile equations using the universal function for momentum φ_m. This results in a new functional relationship depending on the dimensionless parameter $\zeta = z/L$:

$$\varphi_m(\zeta) = \frac{kz}{u_*}\frac{\partial \overline{u}}{\partial z} \tag{4.16}$$

The universal function reflects the impact of mixing. In unstable conditions, enhanced mixing reduces the gradient and increases the flux, while in stable conditions, the opposite occurs. Approximate forms of the universal function, such as the O'KEYPS function, have been given in Sect. 2.4.1. The recommended form is (Högström, 1988):

$$\text{For } \zeta < 0: \quad \varphi_m(\zeta) = (1 - 19.3\zeta)^{-1/4} \tag{4.17}$$

$$\text{For } \zeta > 0: \quad \varphi_m(\zeta) = 1 + 6\zeta \tag{4.18}$$

Figure 4.9 shows the variation of the universal function for momentum with stability parameter according to the equations above.

Integrating the nondimensional wind shear with height, the logarithmic law for non-neutral surface layer is given as:

$$U = \frac{u_*}{k}\ln\left[\frac{z}{z_0} + \psi_m(\zeta)\right] \tag{4.19}$$

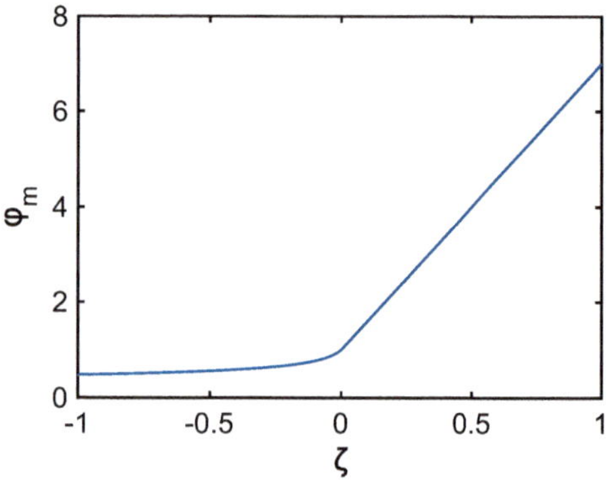

Fig. 4.9 Variation of universal function for momentum with stability parameter

where $\psi_m(\zeta)$ for unstable ($\zeta < 0$) and stable ($\zeta > 0$) conditions is given by:

$$\text{For } \zeta < 0: \quad \psi_m(\zeta) = -2\ln\left(\frac{1+x}{2}\right) - \ln\left(\frac{1+x^2}{2}\right) + 2\tan^{-1}x - \frac{\pi}{2} \tag{4.20}$$

$$\text{For } \zeta > 0: \quad \psi_m(\zeta) = 6\zeta \tag{4.21}$$

where $x = (1 - 19.3\zeta)^{-1/4}$.

Figure 4.10 shows the logarithmic law wind profiles for different atmospheric stability conditions, including unstable ($L = -50$ m), neutral ($L = \infty$), and stable ($L = 50$ m) conditions. The following parameters are assumed: $z_0 = 0.03$ m (open terrain), $u_* = 0.5$ m/s.

Different forms of universal functions have been established over the past 40 years. However, applying these functions presents challenges, particularly in terms of normalization. Additionally, all universal functions need to be recalculated in light of the updated von Kármán constant, which is now set at 0.4, as some earlier functions were based on other values such as 0.35 (Högström, 1988).

Applying the universal function in stable conditions also poses challenges. It is well established that the existing formulation tends to underestimate the turbulent exchange processes. Moreover, it is evident that the universal function should be assumed nearly constant to facilitate z-less scaling. However, the scarcity of measurements has hindered the development of a universally applicable formulation (Foken, 2017).

Similar to the neutral logarithmic law, the profile laws given above are only applicable to the surface layer. Gryning et al. (2007) extended the non-neutral wind profile beyond the surface layer (the model for neutral conditions has been discussed in Sect. 4.1.3). For unstable conditions, the form of nondimensional wind shear $\varphi_m(\zeta) = (1 - 12z/L)^{-1/3}$ is used such that:

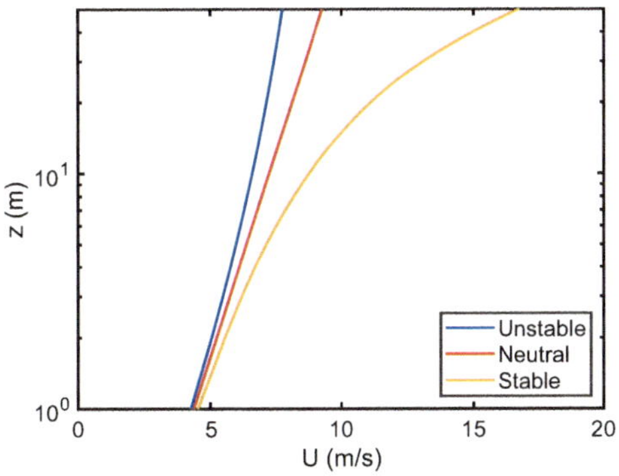

Fig. 4.10 Logarithmic law wind profiles for different atmospheric stability conditions (height axis in logarithmic scale)

$$U = \frac{u_*}{k}\left[\ln\left(\frac{z}{z_0}\right) - \psi_m\left(\frac{z}{L}\right) + \frac{z}{h}\left(1 + \frac{(1-12z/L)^{2/3}-1}{8z/L}\right) + \frac{z}{L_{MBL,U}} - \frac{z}{h}\left(\frac{z}{2L_{MBL,U}}\right)\right] \tag{4.22}$$

$$L_{MBL,U} = \frac{h}{2\left[\left[(\ln(u_*/fz_0) - A)^2 + B^2\right]^{1/2} - \ln(h/z_0) + \psi_m(z/L)\right]} \tag{4.23}$$

$$\psi_m(z/L) = \frac{3}{2}\ln\left(\frac{1+x+x^2}{2}\right) - \sqrt{3}\tan^{-1}\left(\frac{1+2x}{\sqrt{3}}\right) + \frac{\pi}{\sqrt{3}} \tag{4.24}$$

where $x = (1 - 12z/L)^{-1/3}$, and A and B are nondimensional functions of the Kazanski–Monin stability parameter $\mu = u_*/fL$ in the steady state ABL or h/L (where h is the ABL height) in the evolving ABL (from the resistance law introduced in Sect. 2.4.2). For stable conditions:

$$U = \frac{u_*}{k}\left[\ln\left(\frac{z}{z_0}\right) + \frac{bz}{L}\left(1 - \frac{z}{2h}\right) + \frac{z}{L_{MBL,S}} - \frac{z}{h}\left(\frac{z}{2L_{MBL,S}}\right)\right] \tag{4.25}$$

$$L_{MBL,S} = \frac{h}{2\left[\left[(\ln(u_*/fz_0) - A)^2 + B^2\right]^{1/2} - \ln(h/z_0) - bz_0/2L\right]} \tag{4.26}$$

where b is the coefficient from the non-dimensional wind shear $\varphi_m(\zeta) = 1 + b\zeta$, which can be taken as $b = 6$ (Högström, 1988).

Due to the very wide spread of data in empirical plots of A and B versus μ or h/L, the non-neutral wind profile models based on the resistance law were rarely used in practice. Nevertheless, recently these models gathered interest in wind energy studies as such studies need simple methods to simulate the wind profile in the whole ABL.

4.2.2 Wind Profiles in Conventionally Neutral Conditions

As has been introduced in Sect. 2.4.2, truly neutral ABL with $\mu = 0$ and $\mu_N = 0$ (where $\mu_N = N/f$, and $N = \sqrt{(g/\theta_{v0})\partial\theta_v/\partial z}$ is the Brunt–Väisälä frequency in the free atmosphere) is observed only during short transition periods after sunset on a background of residual layers of convective origin. Conventionally neutral ABL, which is developed against a background stable stratification with $\mu = 0$ and $\mu_N > 0$, is much more common.

Therefore, neutral ABLs are almost always conventionally neutral rather than truly neutral, and the effects of the stable background stratification should be taken into consideration. Some state-of-the-art conventionally neutral ABL wind profile models have been developed, e.g., the formula given by Zilitinkevich and Esau (2005):

$$U = \frac{u_*}{k}\left[\ln\left(\frac{z}{z_0}\right) + \left(\frac{C_u}{L_M} - \frac{1}{h}\right)z - \frac{C_u}{2L_M h}z^2\right] \tag{4.27}$$

$$L_M = \frac{u_*}{f\sqrt{1 + (C_M \mu_N)^2}} \tag{4.28}$$

where $C_u = 2.5$ and $C_M = 0.1$. The boundary layer height for conventionally neutral ABL can be estimated as:

$$h = C_h \frac{u_*}{f} \tag{4.29}$$

$$C_h = \left[\frac{1}{C_R^2} + \frac{\mu_N}{C_{CN}^2}\right]^{-1/2} \tag{4.30}$$

where $C_R \cong 0.6$ and $C_{CN} \cong 1.36$.

Examples of wind profiles described by the Zilitinkevich and Esau model are given for a site with the following parameters: $z_0 = 0.03$ m (open terrain), $f = 7.3 \times 10^{-5}\,\text{s}^{-1}$ (latitude of 30°), $u_* = 1$ m/s. The ABL depth is estimated according to Eqs. (4.29) and (4.30) (Fig. 4.11).

Another conventionally neutral ABL wind profile model was developed by Kelly et al. (2019):

$$U = \frac{u_*}{k}\left[\ln\left(\frac{z}{z_0}\right) - \frac{\langle a \rangle}{2}\left(\frac{z}{l_{TD}}\right)^2\right] \tag{4.31}$$

$$l_{TD} = \frac{u_*/N}{\sqrt{-\Pi_{\overline{w\theta 2}} Ro_A^{0.15}}} \tag{4.32}$$

where $\langle a \rangle = -4.3$, $\Pi_{\overline{w\theta 2}} \cong -0.0016$, and $Ro_A = u_*/fh$ is the ABL Rossby number.

Examples of wind profiles described by these models are given for a site with the following parameters: $z_0 = 0.03$ m (open terrain), $f = 7.3 \times 10^{-5}\,\text{s}^{-1}$ (latitude

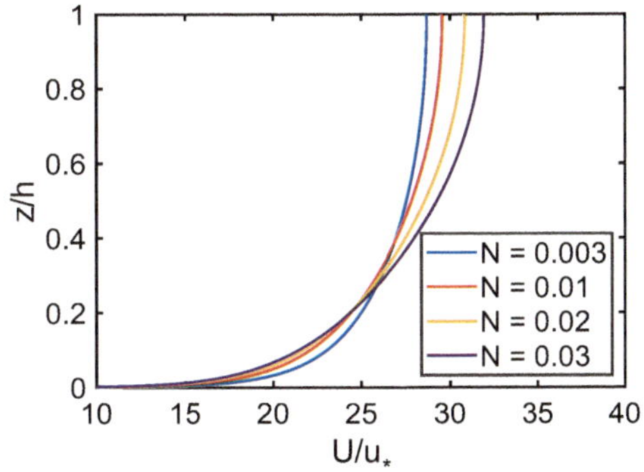

Fig. 4.11 Conventionally neutral wind profiles over open terrain with different Brunt–Väisälä frequency, described by the Zilitinkevich and Esau model

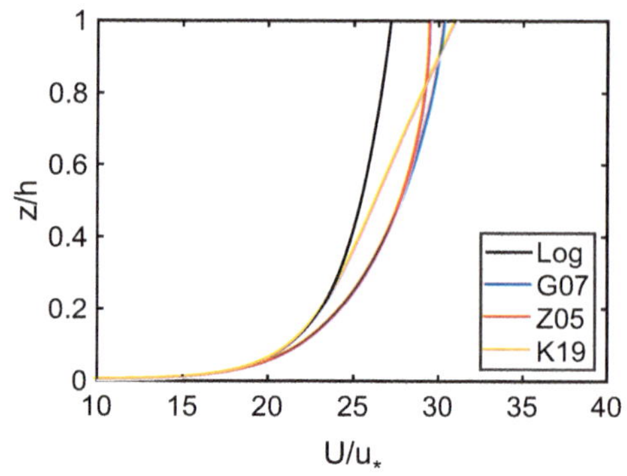

Fig. 4.12 Wind profiles over open terrain described by the logarithmic law, Gryning model, Zilitinkevich and Esau model, and Kelly model

of 30°), and $N = 0.01\,\mathrm{s}^{-1}$. For a storm producing $u_* = 1\,\mathrm{m/s}$, the ABL depth is estimated to be $h \cong 1.6\,\mathrm{km}$ according to Eqs. (4.29) and (4.30). The wind profiles described by the aforementioned models are shown in Fig. 4.12, where Log: Logarithmic; G07: Gryning et al. (2007); Z05: Zilitinkevich and Esau (2005); K19: Kelly et al. (2019).

4.2.3 *Low-Level Jets*

We have focused on stationary conditions arising from various force equilibria. An intriguing non-stationary scenario occurs when frictional forces in the Ekman layer suddenly dissipate during the evening or when winds transition from rough land to smooth sea surfaces. In these situations, the equations that describe the temporal changes in horizontal wind components can be given by:

$$\frac{\partial \overline{u}}{\partial t} = f\left(\overline{v} - v_g\right) \tag{4.33}$$

$$\frac{\partial \overline{v}}{\partial t} = f\left(\overline{u} - u_g\right) \tag{4.34}$$

The left-hand side of these equations incorporates time dependence, allowing for an analytical solution that reveals oscillatory behavior over time:

$$\overline{u} - u_g = D_v \sin ft + D_u \cos ft \tag{4.35}$$

$$\overline{v} - v_g = D_v \cos ft - D_u \sin ft \tag{4.36}$$

D_u and D_v represent the ageostrophic wind components at the onset of the oscillation when friction disappears.

This phenomenon plays a crucial role in the formation of low-level wind maxima, which is usually referred to as "low-level jets (LLJs)". The inertial oscillation is characterized by a rapid increase in wind speed during the initial hours following its onset, accompanied by a continuous change in wind direction until it is disrupted by the mechanical or thermal generation of turbulence after sunrise or after significant travel over the sea surface. In this analytical model, the maximum wind speed can reach twice the geostrophic wind speed, typically occurring after 1/3 pendulum day (approximately seven hours). However, real-world measurements do not align perfectly with this model, noting that maximum wind speeds often occur earlier than predicted, followed by fluctuating flow speeds. The cores of LLJs generally reside at altitudes between 150 and 500 m above ground (Emeis, 2018).

Figure 4.13 illustrates observations of LLJ wind profiles from a radar wind profiler at Cheung Chau Island in Hong Kong. The wind speed shows a rapid increase with height near the surface and peaks at a maximum before declining at higher altitudes. Figure 4.14 provides a summary of the climatology of LLJs observed by the radar wind profiler. Specifically, Fig. 4.14b presents the diurnal variation in the occurrence frequency of LLJ activities. Notably, there is a clear diurnal pattern, with distinct peaks in LLJ occurrence during the early morning (0200–0400 local time) and late evening (2000–2300 local time). These findings indicate that LLJs are primarily nocturnal phenomena, typically developing between midnight and sunrise. The heightened frequency of LLJs during nighttime suggests that diurnal wind oscillations play a significant role in their formation and enhancement.

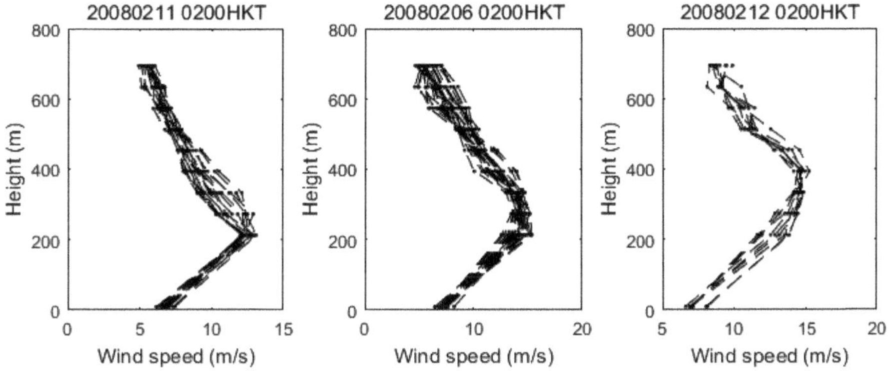

Fig. 4.13 Wind profile observations of low-level jets by a radar wind profiler (Shu et al., 2018)

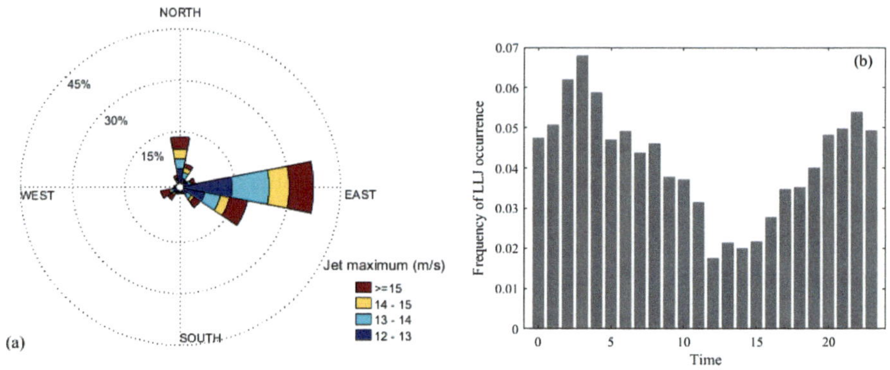

Fig. 4.14 Climatology of low-level jets in Hong Kong observed by a radar wind profiler, **a** distribution of jet speed maximum and jet direction, **b** diurnal variation of occurrence probability, and **c** monthly variation of occurrence probability (Shu et al., 2018)

4.3 Wind Profiles in Storms

4.3.1 Wind Profiles in Tropical Cyclones

Tropical cyclones (TCs) are intense vortical storms that form over warm surface waters in tropical oceans. These systems usually have radial scales of several hundred kilometers, comparable to those of some midlatitude synoptic systems. A mature TC can be divided roughly into three components, the eye, the eyewall, and the spiral rainbands. The eye at the center of the TC is a roughly circular area with a size usually about 30–60 km. Inside the eye, the weather is comparatively fair, wind speeds are low, and the lowest pressure and highest temperature (i.e., warm core) can be found. The eyewall surrounding the eye is composed of a roughly circular ring of dense convective clouds. Conditions there are more severe with the highest wind speed (which may reach 300 km/h or 85 m/s in the most intense TCs) and heavy precipitation. Outside the eye is a long-extended spiral rain bands, which also contain strong wind and heavy rainfall. Information on important aspects of TC fluid dynamics, such as intensification, movement, air-sea interaction, potential intensity, vorticity, and cloud-scale eddies, can be found in Lighthill (1998), Emanuel (2003), Chan (2005), and Montgomery and Smith (2017).

The three-dimensional structure of a mature TC is demonstrated in Fig. 4.15, and the satellite imagery of a very strong TC, i.e., Super Typhoon Mangkhut over the Philippine Sea on 12 September 2018, is shown in Fig. 4.16.

The absolute angular momentum equation for TCs (expressed in cylindrical polar coordinates (r, λ, z), where r is the radius, λ is the azimuthal angle, and z is the height above surface) is given by:

$$\frac{DM}{Dt} = -\frac{1}{\rho}\frac{\partial p}{\partial \lambda} - rF_\lambda \tag{4.37}$$

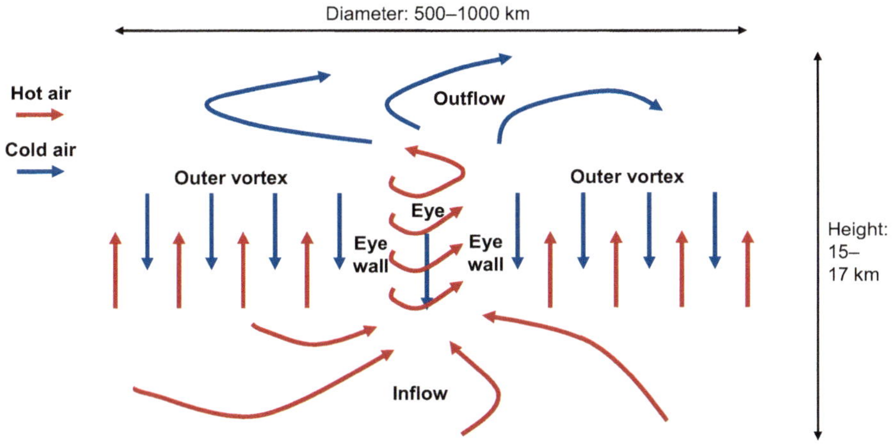

Fig. 4.15 Diagram showing the three-dimensional structure of a mature tropical cyclone

Fig. 4.16 Satellite imagery of a tropical cyclone (imagery courtesy of NASA Earth Observatory)

where $M = rv + (1/2)fr^2$ is the absolute angular momentum per unit mass, F_λ is unsolved processes associated with turbulent momentum diffusion (e.g., the frictional stress and related mixing processes in the ABL). For asymmetric ($\partial/\partial\lambda = 0$) and frictionless ($F_\lambda = 0$) flow, $DM/Dt = 0$, suggesting that M is materially conserved.

However, in the ABL, the frictional stress cannot be neglected. As air parcels converge in the TC ABL, they lose momentum to the surface. If the rate of momentum loss is relatively small (i.e., less than the rate of decrease of radius), the tangential velocity ($v = M/R - 1/2fr^2$) may increase. As a result, at some inner radii of the TC, the tangential wind speed in the ABL may exceed the local value above the ABL. This process is referred to as the boundary layer spin-up mechanism (Montgomery & Smith, 2017; Smith et al., 2009). At such radii, the ABL flow is supergradient, and the maximum tangential wind speed occurs within the ABL rather than above the ABL. This low-level wind maximum or low-level jet (LLJ) phenomenon has been frequently observed in inner cores of TCs (e.g., Kepert, 2006a, 2006b; Powell et al., 2003).

The supergradient flow in TCs makes the wind profiles in the TC ABL different from those in other synoptic-scale strong wind events (e.g., extratropical cyclones and monsoons). There are some empirically derived wind profile models for TC ABL based on Global Position System (GPS) dropsonde observations, e.g., Vickery et al. (2009) constructed the following model:

$$U = \frac{u_*}{k}\left[\ln\left(\frac{z}{z_0}\right) - a\left(\frac{z}{H^*}\right)^n\right] \tag{4.38}$$

where $a = 0.4$, $n = 2.0$, and $H^* = h/1.12$. Figure 4.17 shows the wind profiles described by the logarithmic law and the Vickery model (denoted by V09) with different boundary layer heights.

Snaiki and Wu (2018) proposed the following formula:

$$U = \frac{u_*}{k}\left[\ln\left(\frac{z}{z_0}\right) + \eta_0 \sin\left(\frac{z}{\delta}\right)\exp\left(-\frac{z}{\delta}\right)\right] \tag{4.39}$$

where $\eta_0 = 9.026$, and $\delta = h$.

Figure 4.18a presents normalized composite wind profiles from dropsonde observations in TCs over the South China Sea. The wind speeds are normalized based

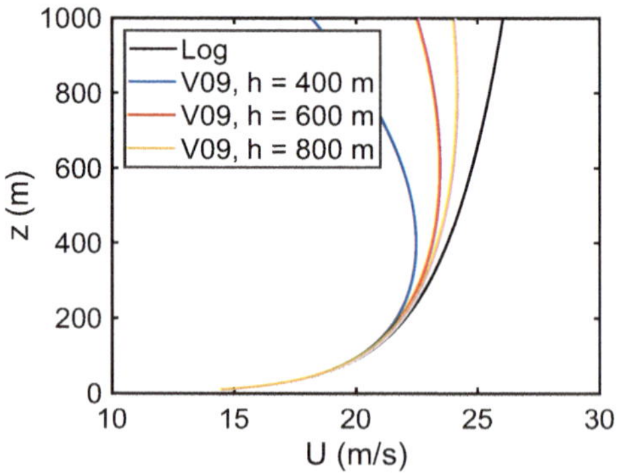

Fig. 4.17 Wind profiles described by the logarithmic law and the Vickery model with different boundary layer heights

on the mean boundary layer (MBL) wind speed, calculated as the average over the height range of 10–500 m (Powell et al., 2003). The wind speed exhibits a logarithmic increase within the lowest 200 m, reaching a peak around 500–700 m before decreasing. This wind profile closely mirrors those recorded by dropsondes in Atlantic TCs (Giammanco et al., 2013; Vickery et al., 2009; Wang et al., 2015). The presence of a logarithmic layer in the lower several hundred meters of TCs is consistent with earlier findings (Franklin et al., 2003; Giammanco et al., 2013). The decline in wind speed above the maximum height can be attributed to strong inward angular momentum advection (Kepert, 2001) and the boundary layer "spin-up" mechanism (Montgomery & Smith, 2017), which leads to "supergradient" winds in the inner cores of TCs. Variability in individual wind speed measurements may be influenced by mesoscale, convective, and turbulent features within TCs, as well as the descent drift of dropsondes (Powell et al., 2003).

Figure 4.18b organizes wind profiles by MBL wind speed, revealing that their behavior aligns with the composite profiles, showcasing a logarithmic regime in the lower few hundred meters. This finding is consistent with characteristics of a neutral surface layer, where surface roughness influences momentum distribution (Powell et al., 2003). The assumption of neutral stability holds at high wind speeds in tropical cyclones, as mechanical turbulence production greatly exceeds buoyancy effects (He et al., 2020b; Stull, 1988; Tsai et al., 2019). Additionally, there is a slight reduction in the height of the maximum wind speed as MBL wind speed increases, likely due to the diminished height of the logarithmic layer in the eyewalls of intense TCs, where wind speeds are particularly elevated (Giammanco et al., 2013; Kepert, 2001; Vickery et al., 2009).

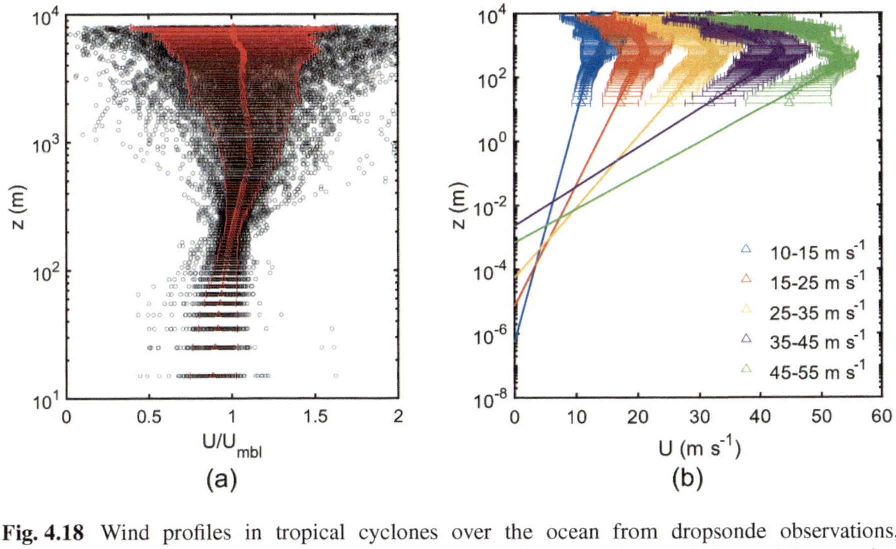

Fig. 4.18 Wind profiles in tropical cyclones over the ocean from dropsonde observations, **a** composite profiles, and **b** profiles grouped by mean boundary layer wind speed (mean wind speed over heights of 10–500 m)

Figure 4.19 shows wind profiles in TCs over a coastal island (Cheung Chau Island) from radar wind profiler observations. As wind speed increases, the jet feature becomes more pronounced, with jet heights generally decreasing as MBL wind speed rises, typically ranging from 500 to 1200 m. The Vickery model is effective in reproducing these jet-like profiles; however, its parameters, particularly jet height, are sensitive to measurement variations. The other two models provide good fits for mean wind speeds below approximately 400 m. Notably, the roughness length z_0 and the power law exponent parameter α increase with wind speed until reaching about 20–25 m/s, after which they decline. However, the critical wind speed associated with the maximum z_0 identified in this study (12–15 m/s at 10 m) is significantly lower than the approximately 33 m/s reported by Powell et al. (2003), likely due to differences in air-sea interaction conditions between shallow coastal waters and deeper oceans. The observed z_0 values range from 0.07 to 0.12 m, and α falls between 0.152 and 0.175. The ratios of 10-m surface wind speed to the maximum wind speed (U_{10}/U_h) range from 0.46 to 0.55, which are notably lower than those typically found over deep ocean.

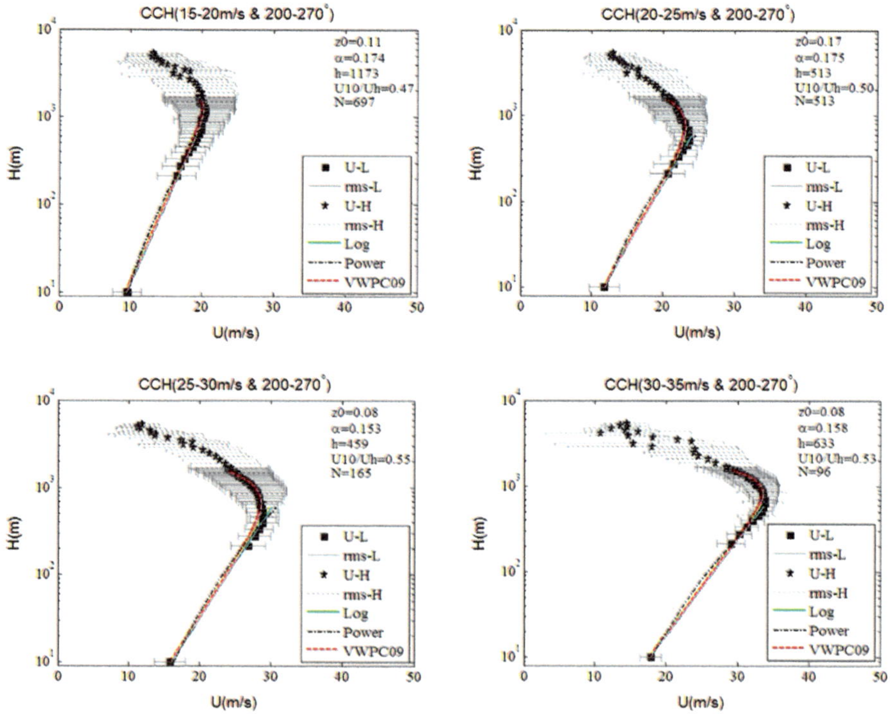

Fig. 4.19 Wind profiles in tropical cyclones over a coastal island from radar wind profiler observations (He et al., 2016)

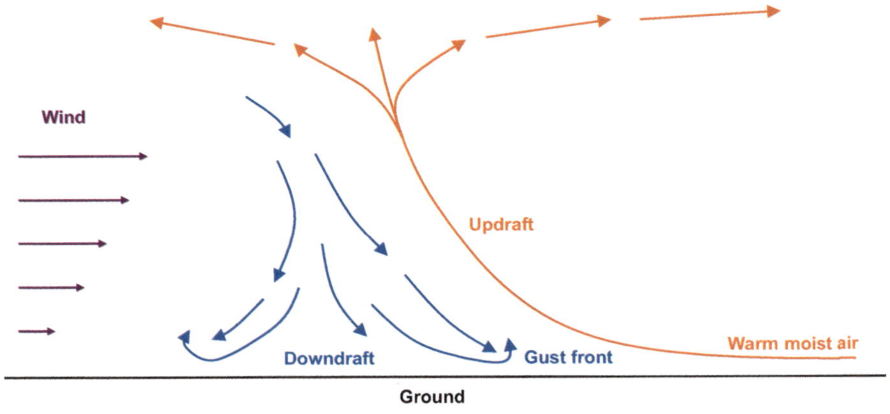

Fig. 4.20 Diagram showing the structure of a thunderstorm

4.3.2 Wind Profiles in Thunderstorms

Thunderstorms (Fig. 4.20) form when warm, moist air rises rapidly in a strong updraft to cooler regions of the atmosphere. As the moisture in the updraft condenses, it forms towering cumulonimbus clouds, which can lead to precipitation. Subsequently, columns of cooled air descend, creating strong downbursts and horizontal winds upon reaching the ground. In some cases, intense thunderstorms may produce swirling air vortices that can intensify and develop into tornadoes.

Thunderstorms have numerous forms, varying in scale from isolated thunderstorms that consist of a single convective cell to mesoscale convective complexes made up of ensembles of multi-celled thunderstorms. Four primary types of thunderstorms can be classified: single-cell, multi-cell, supercell, and squall line (multi-cell line), among which the supercell thunderstorm is usually the most intense. Figure 4.21 shows the weather radar imagery of four squall lines.

Thunderstorms can generate powerful downbursts that result in destructive winds at the surface. These downbursts are classified as macrobursts or microbursts based on their size. A macroburst has a diameter greater than 4 km and can produce ground winds as high as 60 m/s. A microburst is smaller in dimension (diameter smaller than 4 km) but produces winds up to 75 m/s.

Oseguera and Bowles (1988) proposed a useful model for the velocity profiles in both vertical and horizontal directions within a downburst. The horizontal and vertical velocity components are given by:

$$U = \frac{\lambda R^2}{2r}\left\{1 - \exp\left[-\left(\frac{r}{R}\right)^2\right]\right\}\left[\exp\left(-\frac{z}{z^*}\right) - \exp\left(-\frac{z}{\varepsilon}\right)\right] \tag{4.40}$$

$$\overline{w} = -\lambda \exp\left[-\left(\frac{r}{R}\right)^2\right]\left\{\varepsilon\left[\exp\left(-\frac{z}{\varepsilon}\right) - 1\right] - z^*\left[\exp\left(-\frac{z}{z^*}\right) - 1\right]\right\} \tag{4.41}$$

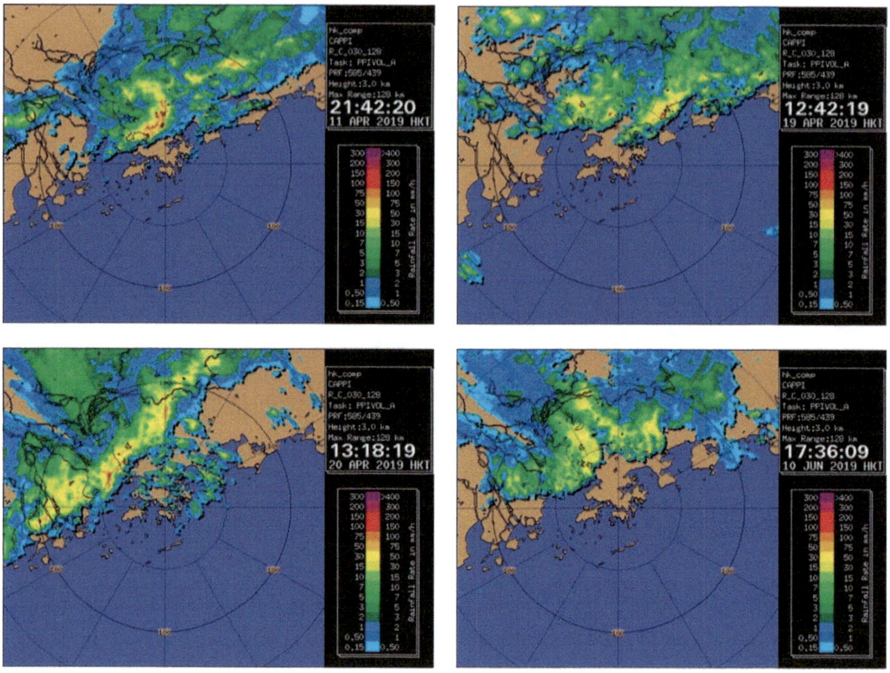

Fig. 4.21 Weather radar imagery of four thunderstorm squall lines (He et al., 2020a)

where r is the radial coordinate from the centre of the downburst, R the characteristic radius of the downburst shaft, z^* is a characteristic height out of the boundary layer, ε is a characteristic height in the boundary layer, and λ is a scaling factor (Oseguera & Bowles, 1988). This model adheres to the principles of fluid mass continuity, although it does not account for the movement of the storm itself.

Figure 4.22 presents examples of wind profiles in thunderstorms, with mean wind speeds normalized by the mean boundary layer (MBL) wind speed. The observation was performed using a radar wind profiler at Cheung Chau Island in Hong Kong. The results reveal clear low-level jet (LLJ) features of thunderstorms at around 400 m, where the normalized wind speed can peak at 1.3. As altitude increases beyond this point, the normalized wind speed gradually decreases, continuing up to approximately 1600 m.

To further illustrate the LLJ characteristics in thunderstorm wind profiles, Fig. 4.23 presents mean wind speed profiles at 13 different time points during a thunderstorm event on April 11, 2019, with a corresponding time history in the upper left panel. During the ramp-up phase, LLJ streams are clearly visible, with the height of maximum wind speed increasing from approximately 50 m at 200 s to 150 m at 290 s. At 290 s, a distinct LLJ is identified, where wind speeds rise from 13 m/s at 10 m to 20 m/s at 150 m, then decrease to 17 m/s at 350 m. These nose-shaped wind profiles have been documented in previous studies (Fujita, 1981; Gunter & Schroeder,

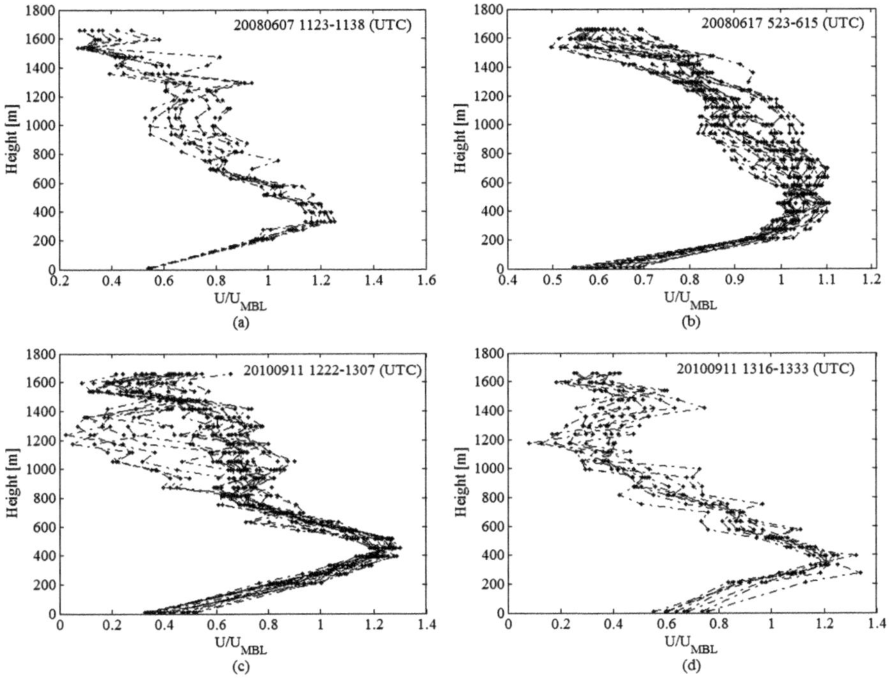

Fig. 4.22 Wind profiles in thunderstorms (Shu et al., 2017)

2015; Shu et al., 2017). Following the peak wind speed, a plateau between 290 and 420 s shows an increase in wind speed with height in the lowest 150 m, remaining relatively constant between 150 and 350 m. This uniform wind speed profile above the maximum height has also been observed in other research (Canepa et al., 2020; Choi, 2004). During the ramp-down phase from 420 to 620 s, wind speed increases with height in the lowest 100 m, remains steady between 100 and 250 m, and then gradually rises again until 350 m.

Figure 4.24 depicts the mean wind speed profiles at 13 time instants for a thunderstorm event on June 10, 2019. The first peak at 440 s shows a nose-shaped profile with a jet height of 160 m, where wind speeds increase from 13.9 m/s at 10 m to 22.5 m/s at 160 m, before slightly decreasing to 21.1 m/s at 350 m. A second peak at 560 s also indicates a low-level wind speed maximum, with a jet height of 300 m, which is higher than that of the first peak. Although wind speeds likely decrease above 350 m at the second peak, the lack of observations above this height prevents confirmation. After the maximum wind, the LLJ features diminish, and wind speeds generally increase with height up to 350 m. The jet heights observed during these events are comparable to those reported in other studies, such as 100–200 m in Choi (2004), 110–350 m in Gunter and Schroeder (2015), and 60–250 m in Canepa et al. (2020), but lower than 400 m in Shu et al. (2017). This variation may be attributed to

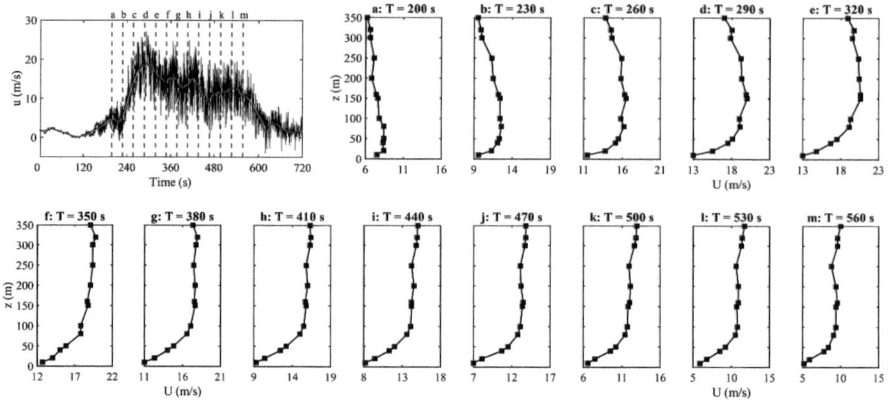

Fig. 4.23 Time-varying wind profiles in a thunderstorm event on 11 April 2019 (He et al., 2021)

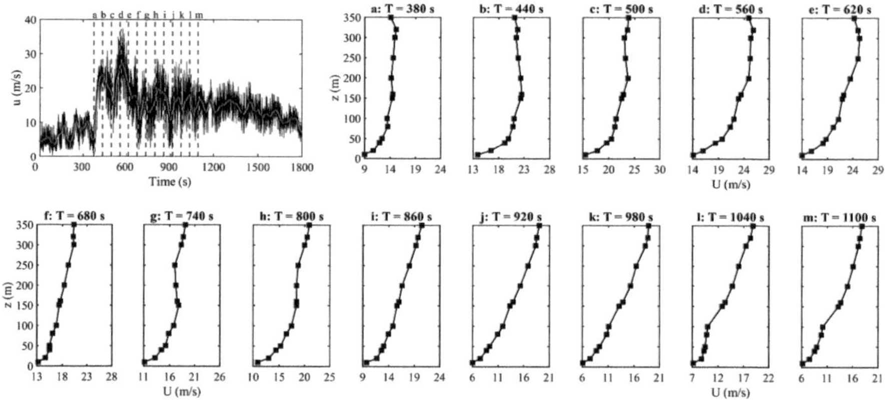

Fig. 4.24 Time-varying wind profiles in a thunderstorm event on 10 June 2019 (He et al., 2021)

differences in the distance from the observation site to the thunderstorm center and the overall scale of the thunderstorms.

Despite both events exhibiting similar wind directions of 240–300°, indicative of a generally flat and open upwind terrain, their wind speed profiles differ significantly. In the April 11 event, a clear LLJ is identified at a height of 150 m, with a power law exponent of 0.14 below this jet height, while wind speeds above the jet height decrease slightly from 20 m/s at 150 m to 17 m/s at 350 m. In contrast, the June 10 event features a jet height of 320 m and a higher power law exponent of 0.17, suggesting that while wind profiles in thunderstorms may have some connection to upwind terrain, they are predominantly influenced by the storm's circulation structures.

References

AIJ. (2015). *AIJ-RLB-2015 recommendations for loads on buildings*. Architecture Institute of Japan.

AS/NZS. (2011). *AS/NZS 1170.2:2011 structural design actions*. Joint Standards Australia/ Standards New Zealand Committee.

ASCE. (2022). *ASCE7-22 minimum design loads and associated criteria for buildings and other structures*. American Society of Civil Engineering.

Barnéoud, P., & Ek, N. (2020). On the application of linear regression to surface-layer wind profiles for deducing roughness length and friction velocity. *Boundary-Layer Meteorology, 174*, 327–339. https://doi.org/10.1007/s10546-019-00479-8

Blackadar, A. K. (1962). The vertical distribution of wind and turbulent exchange in a neutral atmosphere. *Journal of Geophysical Research, 67*, 3095–3102. https://doi.org/10.1029/JZ067i008p03095

Canepa, F., Burlando, M., & Solari, G. (2020). Vertical profile characteristics of thunderstorm outflows. *Journal of Wind Engineering & Industrial Aerodynamics, 206*, 104332. https://doi.org/10.1016/j.jweia.2020.104332

CEN. (2005). *EN 1991-1-4:2005 Eurocode 1: Actions on structures—Part 1–4: General actions–wind actions*. European Committee for Standardization.

Chan, J. C. L. (2005). The physics of tropical cyclone motion. *Annual Review of Fluid Mechanics, 37*(1), 99–128. https://doi.org/10.1146/annurev.fluid.37.061903.175702

Charnock, H. (1955). Wind stress on a water surface. *Quarterly Journal of the Royal Meteorological Society, 81*, 639–640. https://doi.org/10.1002/qj.49708135027

Choi, E. C. C. (2004). Field measurement and experimental study of wind speed profile during thunderstorms. *Journal of Wind Engineering & Industrial Aerodynamics, 92*, 275–290. https://doi.org/10.1016/j.jweia.2003.12.001

CNS. (2012). *GB 50009-2012 load code for the design of building structures*. National Standards Committee of China.

Deaves, D. M., & Harris, R. I. (1978). *A mathematical model of the structure of strong winds* (CIRIA report 76). London, the United Kingdom.

Emanuel, K. (2003). Tropical cyclones. *Annual Review of Earth and Planetary Sciences, 31*(1), 75–104. https://doi.org/10.1146/annurev.earth.31.100901.141259

Emeis, S. (2018). *Wind energy meteorology—Atmospheric physics for wind power generation* (2nd ed.). Springer.

Foken, T. (2017). *Micrometeorology* (2nd ed.). Springer.

Franklin, J. L., Black, M. L., & Valde, K. (2003). GPS dropwindsonde wind profiles in hurricanes and their operational implications. *Weather and Forecasting, 18*, 32–44. https://doi.org/10.1175/1520-0434(2003)018<0032:GDWPIH>2.0.CO;2

Fujita, T. T. (1981). Tornadoes and downbursts in the context of generalized planetary scales. *Journal of the Atmospheric Sciences, 38*, 1511–1534.

Giammanco, I. M., Schroeder, J. L., Masters, F. J., Vickery, P. J., Krupar, R. J., & Balderrama, J. A. (2016). Influences on observed near-surface gust factors in landfalling U.S. Gulf coast hurricanes: 2004–08. *Journal of Applied Meteorology and Climatology, 55*, 2587–2611. https://doi.org/10.1175/JAMC-D-16-0053.1

Giammanco, I. M., Schroeder, J. L., & Powell, M. D. (2013). GPS dropwindsonde and WSR-88D observations of tropical cyclone vertical wind profiles and their characteristics. *Weather and Forecasting, 28*, 77–99. https://doi.org/10.1175/WAF-D-11-00155.1

Grimmond, C. S. B., King, T. S., Roth, M., & Oke, T. R. (1998). Aerodynamic roughness of urban areas derived from wind observations. *Boundary-Layer Meteorology, 89*, 1–24. https://doi.org/10.1023/A:1001525622213

Gryning, S. E., Batchvarova, E., Brümmer, B., Jørgensen, H., & Larsen, S. (2007). On the extension of the wind profile over homogeneous terrain beyond the surface boundary layer. *Boundary-Layer Meteorology, 124*, 251–268. https://doi.org/10.1007/s10546-007-9166-9

Gunter, W. S., & Schroeder, J. L. (2015). High-resolution full-scale measurements of thunderstorm outflow winds. *Journal of Wind Engineering & Industrial Aerodynamics, 138*, 13–26. https://doi.org/10.1016/j.jweia.2014.12.005

He, J. Y., Chan, P. W., Li, Q. S., Li, L., Lu, C., Zhang, L., & Yang, H. L. (2020a). Characteristics of wind structure and nowcasting of gust associated with subtropical squall lines over Hong Kong and Shenzhen, China. *Atmosphere, 11*, 270. https://doi.org/10.3390/atmos11030270

He, J. Y., He, Y. C., Li, Q. S., Chan, P. W., Zhang, L., Yang, H. L., & Li, L. (2020b). Observational study of wind characteristics, wind speed and turbulence profiles during Super Typhoon Mangkhut. *Journal of Wind Engineering & Industrial Aerodynamics, 206*, 104362. https://doi.org/10.1016/j.jweia.2020.104362

He, J. Y., Li, Q. S., Chan, P. W., Li, L., Lu, C., Zhang, L., & Yang, H. L. (2021). Characteristics and vertical profiles of mean wind and turbulence for typhoon, monsoon, and thunderstorm winds. *Journal of Structural Engineering, 147*, 04021188. https://doi.org/10.1061/(asce)st.1943-541x.0003156

He, Y. C., Chan, P. W., & Li, Q. S. (2016). Observations of vertical wind profiles of tropical cyclones at coastal areas. *Journal of Wind Engineering & Industrial Aerodynamics, 152*, 1–14. https://doi.org/10.1016/j.jweia.2016.01.009

Högström, U. (1988). Non-dimensional wind and temperature profiles in the atmospheric surface layer: A re-evaluation. *Boundary-Layer Meteorology, 42*, 55–78. https://doi.org/10.1007/BF00119875

ISO. (2009). *ISO 4354:2009 wind actions on structures*. International Organization for Standardization.

Kelly, M., Cersosimo, R. A., & Berg, J. (2019). A universal wind profile for the inversion-capped neutral atmospheric boundary layer. *Quarterly Journal of the Royal Meteorological Society, 145*, 982–992. https://doi.org/10.1002/qj.3472

Kepert, J. (2001). The dynamics of boundary layer jets within the tropical cyclone core. Part I: Linear theory. *Journal of the Atmospheric Sciences, 58*, 2469–2484. https://doi.org/10.1175/1520-0469(2001)058<2469:TDOBLJ>2.0.CO;2

Kepert, J. D. (2006a). Observed boundary layer wind structure and balance in the hurricane core. Part I: Hurricane Georges. *Journal of the Atmospheric Sciences, 63*, 2169–2193. https://doi.org/10.1175/JAS3745.1

Kepert, J. D. (2006b). Observed boundary layer wind structure and balance in the hurricane core. Part II: Hurricane Mitch. *Journal of the Atmospheric Sciences, 63*, 2194–2211. https://doi.org/10.1175/JAS3746.1

Kwon, D. K., & Kareem, A. (2013). Comparative study of major international wind codes and standards for wind effects on tall buildings. *Engineering Structures, 51*, 23–35. https://doi.org/10.1016/j.engstruct.2013.01.008

Lettau, H. (1969). Note on aerodynamic roughness-parameter estimation on the basis of roughness-element description. *Journal of Applied Meteorology and Climatology, 8*(5), 828–832. https://doi.org/10.1175/1520-0450(1969)008<0828:NOARPE>2.0.CO;2

Lighthill, J. (1998). Fluid mechanics of tropical cyclones. *Theoretical and Computational Fluid Dynamics, 10*, 3–21. https://doi.org/10.1007/s001620050048

Monin, A. S., & Obukhov, A. M. (1954). Osnovnye zakonomernosti turbulentnogo peremeshivanija v prizemnom sloe atmosfery (Basic laws of turbulent mixing in the atmosphere near the ground). *Trudy Geofizicheskogo Instituta Akademiya Nauk SSSR, 24*, 163–187.

Montgomery, M. T., & Smith, R. K. (2017). Recent developments in the fluid dynamics of tropical cyclones. *Annual Review of Fluid Mechanics, 49*(1), 541–574. https://doi.org/10.1146/annurev-fluid-010816-060022

NRC. (2015). *NRCC 56190 national building code of Canada 2015*. National Research Council Canada.

Oseguera, R. M., & Bowles, R. L. (1988). *A simple analytic 3-dimensional downburst model based on boundary layer stagnation flow* (NASA-TM-100632).

Panofsky, H. A., & Dutton, J. A. (1984). *Atmospheric turbulence, models and methods for engineering applications* (1st ed.). Wiley.

Powell, M. D., Vickery, P. J., & Reinhold, T. A. (2003). Reduced drag coefficient for high wind speeds in tropical cyclones. *Nature, 402*, 279–283. https://doi.org/10.1038/nature01481

Shu, Z. R., Li, Q. S., He, Y. C., & Chan, P. W. (2017). Vertical wind profiles for typhoon, monsoon and thunderstorm winds. *Journal of Wind Engineering & Industrial Aerodynamics, 168*, 190–199. https://doi.org/10.1016/j.jweia.2017.06.004

Shu, Z. R., Li, Q. S., He, Y. C., & Chan, P. W. (2018). Investigation of low-level jet characteristics based on wind profiler observations. *Journal of Wind Engineering & Industrial Aerodynamics, 174*, 369–381. https://doi.org/10.1016/j.jweia.2018.01.035

Smith, R. K., Montgomery, M. T., & Van Sang, N. (2009). Tropical cyclone spin-up revisited. *Quarterly Journal of the Royal Meteorological Society, 135*(642), 1321–1335. https://doi.org/10.1002/qj.428

Snaiki, R., & Wu, T. (2018). A semi-empirical model for mean wind velocity profile of landfalling hurricane boundary layers. *Journal of Wind Engineering & Industrial Aerodynamics, 180*, 249–261. https://doi.org/10.1016/j.jweia.2018.08.004

Stull, R. B. (1988). *An introduction to boundary layer meteorology* (1st ed.). Kluwer Academic Publishers.

Tsai, Y., Miau, J., Yu, C., & Chang, W. (2019). Lidar observations of the typhoon boundary layer within the outer rainbands. *Boundary-Layer Meteorology, 171*, 237–255. https://doi.org/10.1007/s10546-019-00427-6

Vickery, P. J., Wadhera, D., Powell, M. D., & Chen, Y. (2009). A hurricane boundary layer and wind field model for use in engineering applications. *Journal of Applied Meteorology and Climatology, 48*, 381–405. https://doi.org/10.1175/2008JAMC1841.1

Wang, J., Young, K., Hock, T., Lauritsen, D., Behringer, D., Black, M., Black, P. G., Franklin, J., Halverson, J., Molinari, J., Nguyen, L., Reale, T., Smith, J., Sun, B., Wang, Q., & Zhang, J. A. (2015). A long-term, high-quality, high-vertical-resolution GPS dropsonde dataset for hurricane and other studies. *Bulletin of the American Meteorological Society, 96*, 961–973. https://doi.org/10.1175/BAMS-D-13-00203.1

Wieringa, J. (1992). Updating the Davenport roughness classification. *Journal of Wind Engineering & Industrial Aerodynamics, 41*, 357–368. https://doi.org/10.1016/0167-6105(92)90434-C

Zilitinkevich, S. S. (1969). On the computation of the basic parameters of the interaction between the atmosphere and the ocean. *Tellus, 21*.

Zilitinkevich, S. S., & Esau, I. N. (2002). On integral measures of the neutral barotropic planetary boundary layer. *Boundary-Layer Meteorology, 104*, 371–379. https://doi.org/10.1023/A:1016540808958

Zilitinkevich, S. S., & Esau, I. N. (2005). Resistance and heat-transfer laws for stable and neutral planetary boundary layers: Old theory advanced and re-evaluated. *Quarterly Journal of the Royal Meteorological Society, 131*, 1863–1892. https://doi.org/10.1256/qj.04.143

Chapter 5
Turbulence Over Homogenous Terrain

Further to the mean wind profiles discussed in Chap. 4, this chapter outlines the current understanding of turbulence characteristics, e.g., turbulence intensity, gust factors, turbulence integral length scales, and power spectral density, in the atmospheric boundary layer over homogeneous terrain. It begins with an overview of turbulence characteristics in neutral atmospheric stability conditions (Sect. 5.1), followed by a brief introduction to those in non-neutral conditions (Sect. 5.2). Subsequently, recent observations of turbulence characteristics in storms, including tropical cyclones and thunderstorms, are discussed (Sect. 5.3).

5.1 Turbulence in Neutral Conditions

Atmospheric turbulence impacts the design of buildings and structures in three key ways. First, it affects the conversion of wind speeds across various averaging intervals, which in turn influences how design wind speeds are defined in wind codes and standards. Second, turbulence significantly alters the wind flow around structures, affecting the aerodynamic pressures they experience. Third, turbulence can induce substantial resonant dynamic effects, which must be carefully considered when designing flexible structures.

5.1.1 Turbulence Intensity

Turbulence intensity is the ratio of the standard deviation of the wind velocity fluctuating components to the mean wind speed:

$$I_i = \frac{\sigma_i}{U}, \quad i = u, v, w \tag{5.1}$$

J. He and Q.-S. Li, *Winds in the Atmospheric Boundary Layer*,
https://doi.org/10.1007/978-981-95-1517-2_5

where I_u, I_v, and I_w are longitudinal (along-wind), lateral (cross-wind), and vertical turbulence intensities, respectively, and σ_u, σ_v, and σ_w are standard deviations of wind components in longitudinal, lateral, and vertical directions, respectively.

Denoting $\sqrt{\beta_i} = \sigma_i/u_*$, $i = u, v, w$ (which are the integral turbulence characteristics introduced in Sect. 2.4.3), then $I_i = \sqrt{\beta_i}u_*/U$. From this relationship, it is clear that if u_*/U is given by a wind profile model (e.g., logarithmic law in the surface layer), then the turbulence intensity profile can be determined based on $\sqrt{\beta_i}$ (or σ_i/u_*). In the neutral surface layer, values of $\sigma_u/u_* = 2.5$, $\sigma_v/u_* = 2.0$, and $\sigma_w/u_* = 1.25$ were commonly used according to the flux-variation similarity introduced in Sect. 2.4.3. Taking $\sqrt{\beta_u} = \sigma_u/u_* = 2.5$ and $k = 0.4$, the longitudinal turbulence intensity in the neutral surface layer can therefore be expressed as:

$$I_u = \frac{\sqrt{\beta_i}u_*}{U} = \frac{\sqrt{\beta_i}u_*k}{u_*\ln(z/z_0)} = \frac{\sqrt{\beta_i}k}{\ln(z/z_0)} = \frac{1}{\ln(z/z_0)} \tag{5.2}$$

which is the logarithmic form of longitudinal turbulence intensity profile stipulated in EU and ISO. Figure 5.1 shows the longitudinal turbulence intensity profiles in the neutral surface layer with different roughness lengths described by the equation above.

Other wind codes and standards (e.g., ASCE, AIJ, CNS, NBCC, and AS/NZS) generally adopt the empirical power law form of longitudinal turbulence intensity profile:

$$I_u = c\left(\frac{10}{z}\right)^d \tag{5.3}$$

where c is the turbulence intensity factor associated with the magnitude of I_u at 10 m, and d is the turbulence intensity power law exponent governing the shape of the profile.

The values of c and d (or z_0) stipulated in major wind codes and standards for different exposure categories are summarized in Table 5.1. The classification of

Fig. 5.1 Longitudinal turbulence intensity profiles in the neutral surface layer with different roughness lengths

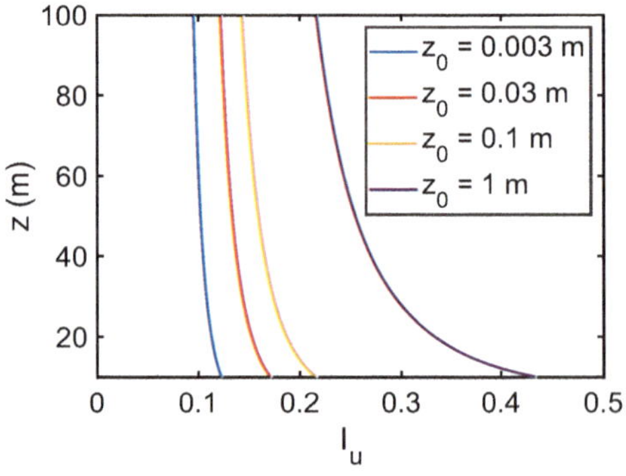

exposure categories is the same as that in Sect. 4.1.4, i.e., I: city center with heavy concentration of tall buildings; II: general urban area; III: suburban area; IV: open terrain area where basic wind velocity is defined; V: open lake or land with minimal obstruction; and VI: sea or coastal area (Kwon & Kareem, 2013).

Figure 5.2 shows the longitudinal turbulence intensity profiles for different terrain categories suggested by the Chinese Load Code for the Design of Building Structures. Note that the turbulence intensity above the gradient heights, which are 300, 350, 450, and 550 m for sea, country, city, and city center, respectively, is not stipulated in the code.

In addition to $\sigma_u/u_* = 2.5$, other values of σ_u/u_* have been suggested in the literature, resulting in different estimates of turbulence intensity. Solari and Piccardo (2001) considered the observed decrease of σ_u/u_* with increasing surface roughness

Table 5.1 Coefficients for longitudinal turbulence intensity profiles in major wind codes and standards

Exposure category		ASCE	AIJ	CNS	NBCC	AS/NZS	EU	ISO
I	c	–	0.53	–	–	0.45	–	z_0 (m) = 3
	d		0.4			0.3		
II	c	–	0.36	0.39	0.51	–	z_0 (m) = 1	–
	d		0.32	0.3	0.36			
III	c	0.3	0.26	0.23	0.27	0.32	z_0 (m) = 0.3	z_0 (m) = 0.3
	d	0.167	0.25	0.22	0.25	0.3		
IV	c	0.2	0.20	0.14	0.16	0.26	z_0 (m) = 0.05	z_0 (m) = 0.03
	d	0.167	0.2	0.15	0.14	0.3		
V	c	0.15	0.16	0.12	–	0.19	z_0 (m) = 0.01	z_0 (m) = 0.003
	d	0.167	0.15	0.12		0.3		
VI	c	–	–	–	–	–	z_0 (m) = 0.003	–
	d							

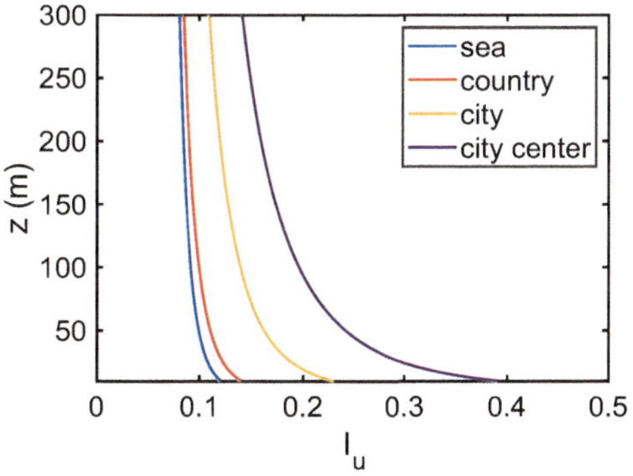

Fig. 5.2 Longitudinal turbulence intensity profiles for different terrain categories stipulated in the Chinese Load Code for the Design of Building Structures

and suggested that in the neutral surface layer:

$$\frac{\sigma_u}{u_*} = \sqrt{6 - 1.1\tan^{-1}(\ln z_0 + 1.75)} \tag{5.4}$$

where z_0 is the roughness length in m.

Simiu and Yeo (2019) also considered the dependence of σ_u/u_* on surface roughness and suggested that in the surface layer $(\sigma_u/u_*)^2 = \beta_{u,s} = 6.5$, 6, 5.25, and 4.9 for $z_0 = 0.005$, 0.03, 0.3, and 1 m, respectively. They further recommended that at greater heights:

$$\beta_u = \beta_{u,s}\exp\left(-1.5\frac{z}{h}\right) \tag{5.5}$$

ESDU (2001) recommended the following form of σ_u/u_*:

$$\frac{\sigma_u}{u_*} = \frac{7.5\eta[0.538 + 0.09\ln(z/z_0)]^{\eta^{16}}}{1 + 0.156\ln(u_*/fz_0)} \tag{5.6}$$

$$\eta = 1 - \frac{z}{h} \tag{5.7}$$

where h can be taken as $h = u_*/6f$.

Furthermore, considering the blocking and shear in the eddy surface layer, which is the lowest part of the surface layer with a height of approximately 20–30 m, Högström et al. (2002) proposed the following form of σ_u/u_*:

$$\frac{\sigma_u}{u_*} = \sqrt{2.5 + \ln\frac{0.46u_*Ak}{zf}} \tag{5.8}$$

where A is a parameter that increases with height.

The lateral and vertical turbulence intensities are also important in some applications. Solari and Piccardo (2001) summarized the measurements of the ratios σ_v/σ_u ($= I_v/I_u$) and σ_w/σ_u ($= I_w/I_u$) in the neutral surface layer, which leads to the estimates $\sigma_v/\sigma_u \cong 0.75$ and $\sigma_w/\sigma_u \cong 0.5$, while Holmes (2015) used $\sigma_v/\sigma_u \cong 0.88$ and $\sigma_w/\sigma_u \cong 0.55$. ESDU (2001) gives the following expressions:

$$\frac{I_v}{I_u} = \frac{\sigma_v}{\sigma_u} = 1 - 0.22\cos^4\left(\frac{\pi}{2}\frac{z}{h}\right) \tag{5.9}$$

$$\frac{I_w}{I_u} = \frac{\sigma_w}{\sigma_u} = 1 - 0.45\cos^4\left(\frac{\pi}{2}\frac{z}{h}\right) \tag{5.10}$$

where h can be taken as $h = u_*/6f$.

5.1.2 Gust Factor

In many design codes and standards for wind loading, a peak gust wind speed is used for design purposes. Given the instantaneous wind velocity $u = \overline{u} + u'$, the peak gust wind speed $\hat{u}(t, T)$ can be defined as the maximum value of u averaged over a short period t (generally taken as 3 s) and occurring in a long observation period T (usually taken as 600 or 3600 s). The gust factor is the ratio of the maximum gust speed to the mean wind speed within the observation period:

$$G_u(t, T) = \frac{\hat{u}(t)}{U(T)} \tag{5.11}$$

$G_v(t, T)$ and $G_w(t, T)$ can be defined in the same way. Assuming that the wind velocity components follow the Gaussian distribution, it can be shown that the expected peak gust is given approximately by (here shown for the longitudinal wind velocity component):

$$\hat{u}(t) = U(T) + g_u(t, T)\sigma_u(T) \tag{5.12}$$

where $g_u(t, T)$ is the peak factor. Substituting Eq. (5.12) into Eq. (5.11), the gust factor is then expressed as:

$$G_u(t, T) = 1 + g_u(t, T)\frac{\sigma_u(T)}{U(T)} = 1 + g_u(t, T)I_u(T) \tag{5.13}$$

To approximate the peak factor, ESDU (2012) gives the following formulae:

$$g_u(t, T) = \left(\sqrt{2\ln[T\nu(t, T)]} + \frac{\gamma}{\sqrt{2\ln[T\nu(t, T)]}}\right)\frac{\sigma_u(t, T)}{\sigma_u} \tag{5.14}$$

$$\nu(t, T)^2 = \frac{\int_0^\infty n^2 S_u(n)\chi^2(n, t, T)dn}{\int_0^\infty S_u(n)\chi^2(n, t, T)dn} \tag{5.15}$$

$$\chi^2(n, t, T) = \left[\frac{\sin(\pi nt)}{\pi nt}\right]^2 - \left[\frac{\sin(\pi nT)}{\pi nT}\right]^2 \tag{5.16}$$

where $S_u(n)$ can be approximated by the von Kármán spectrum (to be discussed in Sect. 2.6.5) as:

$$\frac{nS_u(n)}{\sigma_u^2} = \frac{4nT_u}{[1 + 70.8(nT_u)]^{5/6}} \tag{5.17}$$

$$T_u = 3.13z^{0.2} \tag{5.18}$$

For $T = 3600\,\text{s}$ (1 h), $\sigma_u(t, T)/\sigma_u$ and $\nu(t, T)$ can be approximated by:

$$\frac{\sigma_u(t, 3600\,\mathrm{s})}{\sigma_u} = 1 - 0.193\left[\frac{T_u}{t} + 0.1\right]^{-0.68} \tag{5.19}$$

$$\nu(t, 3600\,\mathrm{s}) = \frac{0.007 + 0.213(T_u/t)^{0.654}}{T_u} \tag{5.20}$$

Figure 5.3 shows variation of longitudinal gust factor with gust duration described by the ESDU model and the Durst curve (Durst, 1960), assuming $T = 3600\,\mathrm{s}$, $z = 10\,\mathrm{m}$, $U = 10\,\mathrm{m/s}$, and $f = 7.3 \times 10^{-5}\,\mathrm{s}^{-1}$ (latitude of 30°).

Figure 5.4 shows longitudinal gust factor profiles with different roughness lengths and displacement heights described by the ESDU model, assuming $t = 3\,\mathrm{s}$, $T = 3600\,\mathrm{s}$, $z_0 = 0.03\,\mathrm{m}$ (open terrain), $f = 7.3 \times 10^{-5}\,\mathrm{s}^{-1}$ (latitude of 30°), and $u_* = 1\,\mathrm{m/s}$.

In addition to the peak factor approach introduced above, there are other relationships for empirically deriving the longitudinal gust factor from the turbulence intensity. One of the most cited relationships is given as (Cao et al., 2009):

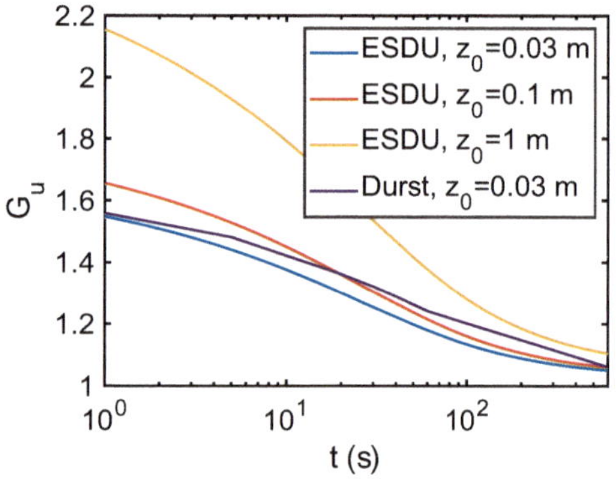

Fig. 5.3 Variation of longitudinal gust factor with gust duration

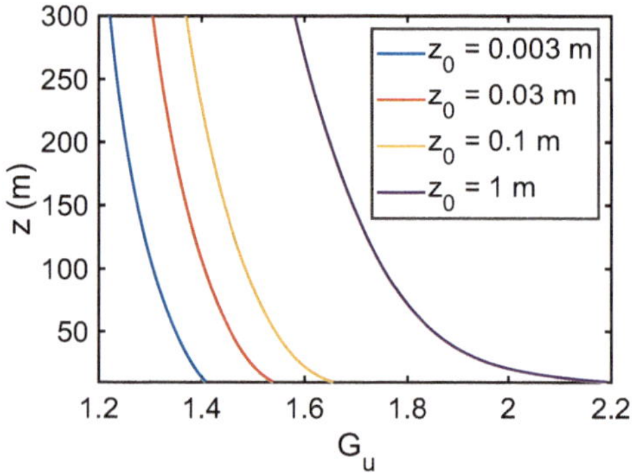

Fig. 5.4 Longitudinal gust factor profiles for different roughness lengths described by the ESDU model

$$G_u(t,T) = 1 + k_1 \times I_u^{k_2} \ln\left(\frac{T}{t}\right) \tag{5.21}$$

where k_1 and k_2 are two empirical coefficients. Ishizaki (1983) suggested $k_1 = 0.5$ and $k_2 = 1$, Choi (1978) recommended $k_1 = 0.62$ and $k_2 = 1.27$, and Cao et al. (2009) found $k_1 = 0.5$ and $k_2 = 1.15$. The variation of longitudinal gust factor with turbulence intensity described by this model is shown in Fig. 5.5, where C78: Choi (1978); I83: Ishizaki (1983); C09: Cao et al. (2009); and peak: Eq. (5.13) with a peak factor of 3.7.

In case the wind velocity is non-Gaussian, Kareem and Zhao (1994) proposed the following expressions for the non-Gaussian peak factor based on a moment-based Hermite model:

$$g(t,T) = \kappa\left[\mu + h_3\left(\mu^2 - 1\right) + h_4\left(\mu^3 - 3\mu\right)\right] \tag{5.22}$$

$$\kappa = \frac{1}{\sqrt{1 + 2h_3^2 + 6h_4^2}} \tag{5.23}$$

$$\mu = \sqrt{2\ln[T\nu(t,T)]} + \frac{\gamma}{\sqrt{2\ln[T\nu(t,T)]}} \tag{5.24}$$

$$h_3 = \frac{\gamma_3}{4 + 2\sqrt{1 + 1.5(\gamma_4 - 3)}} \tag{5.25}$$

$$h_4 = \frac{\sqrt{1 + 1.5(\gamma_4 - 3)} - 1}{18} \tag{5.26}$$

where γ_3 and γ_4 are the skewness and kurtosis of the wind velocity component, respectively.

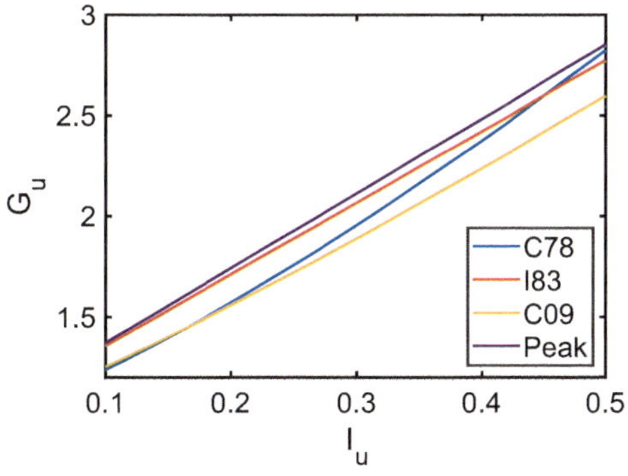

Fig. 5.5 Variation of longitudinal gust factor with turbulence intensity

5.1.3 Turbulence Integral Length Scale

Turbulence integral length scale is a measure of the average size of energy-containing eddies in the flow, with the larger eddies governing mainly its magnitude. From the viewpoint of structural engineering, the turbulence integral length scale controls to what extent the turbulence components of wind flow will engulf a structure. Small values of turbulence integral length scales relative to the dimensions of the structure suggest a relatively insignificant influence of the velocity fluctuations on the overall wind loading. On the contrary, if turbulence integral length scales are large, the eddy will envelop the whole structure, and the influence will be significant (Simiu & Yeo, 2019).

Turbulence integral length scale has nine components related to the fluctuating component of the wind velocity and the spatial extent of the eddy, which can be expressed by L_i^j, $i = u, v, w, j = x, y, z$. For instance, the scale L_w^x is a measure of the longitudinal spatial extent (x) of the vertical fluctuating velocity component (w). The turbulence integral length scale is defined as:

$$L_i^j = \int_0^\infty \frac{1}{U^2} R_{ii}^j(r) dr, \quad i = u, v, w, \ j = x, y, z \tag{5.27}$$

where $R_{ii}^j(r)$ is the autocorrelation function of the wind component $i = u, v, w$ in the $j = x, y, z$ direction. For the observation point $\mathbf{x}$ and distance vector $\mathbf{r}$ (in the direction of j, e.g., $\mathbf{r} = (r, 0, 0)$ for $j = x$) at time t, the autocorrelation function is given as:

$$R_{ii}^j(\mathbf{x}, \mathbf{r}, t) = \overline{u_i'(\mathbf{x}, t) u_i'(\mathbf{x} + \mathbf{r}, t)}, \quad i = u, v, w, \ j = x, y, z \tag{5.28}$$

According to Taylor's frozen turbulence hypothesis, L_i^x defined in Eq. (5.27) in Lagrangian terms can alternatively be defined in Eulerian terms as:

$$L_i^x = U \int_0^\infty \frac{1}{U^2} R_{ii}^x(\tau) d\tau, \quad i = u, v, w \tag{5.29}$$

$$R_{ii}^x(\tau) = \overline{i'(t) i'(t + \tau)}, \quad i = u, v, w \tag{5.30}$$

If not specified, L_i usually refers to L_i^x.

The following empirical formula was suggested by Counihan (1975) for L_u^x in the height range of 10–240 m:

$$L_u^x = C z^{1/n} \tag{5.31}$$

where the coefficients C and $1/n$ are given in Fig. 13 of Counihan (1975).

Simiu and Yeo (2019) showed that in the neutral surface layer, L_u^x can be expressed as:

$$L_u^x = 0.3(\sigma_u/u_*)^3 z \tag{5.32}$$

Solari and Piccardo (2001) considered the dependence of turbulence integral length scales on surface roughness and recommended the following expression:

$$L_u^x = 300\lambda_u^x \left(\frac{z}{200}\right)^{\nu} \tag{5.33}$$

$$\nu = 0.67 + 0.05\ln z_0 \tag{5.34}$$

where $\lambda_u^x = 1$ m. This expression is adopted in Eurocode (EU).

ESDU (2001) considered the variation of turbulence integral length scale with surface roughness and ABL height and gives the following model of L_u^x:

$$L_u^x = \frac{A^{3/2}(\sigma_u/u_*)^3 z}{2.5K_z^{3/2}(1 - z/h)^2(1 + 5.75z/h)} \tag{5.35}$$

$$A = 0.115\left[1 + 0.315\left(1 - \frac{z}{h}\right)^6\right]^{2/3} \tag{5.36}$$

$$K_z = K_\infty\sqrt{1 - \left(1 - \frac{z}{z_c}\right)^2} \tag{5.37}$$

$$z_c = 0.39h\left(\frac{u_*}{fz_0}\right)^{-1/8} \tag{5.38}$$

where $K_\infty = 0.188$, σ_u/u_* is given by Eq. (5.6), and h can be taken as $h = u_*/6f$.

Furthermore, considering the properties of the eddy surface layer, Simiu et al. (2019) proposed the following expression of L_u^x:

$$L_u^x = \left[\frac{0.26f_s^{-5/3}}{4(\sigma_u/u_*)^2}\exp\frac{(\sigma_u/u_*)^2 - 0.65f_s^{-2/3}}{0.26f_s^{-5/3}}\right]z \tag{5.39}$$

where $f_s = 0.125$.

In wind codes and standards, the empirical power law form of longitudinal turbulence integral length scale profile is adopted:

$$L_u^x = l\left(\frac{z}{10}\right)^{\varepsilon} \tag{5.40}$$

where l is the integral length scale factor corresponding to the magnitude of L_u^x at 10 m, and ε is the integral length scale power law exponent regulating the shape of the profile.

The values of l and ε stipulated in major wind codes and standards for different exposure categories are summarized in Table 5.2 (Kwon & Kareem, 2013). The classification of exposure categories is introduced in Sect. 5.1.1.

Figure 5.6 shows the longitudinal turbulence integral length scale profiles for general urban area stipulated in different codes and standards.

In terms of the lateral and vertical turbulence integral length scales, Solari and Piccardo (2001) suggested $L_v^x = 0.25L_u^x$ and $L_w^x = 0.1L_u^x$, while ESDU (2001) recommended that:

Table 5.2 Coefficients for longitudinal turbulence integral length scale profiles in major wind codes and standards

Exposure category		ASCE	AIJ	AS/NZS	EU	ISO
I	l (m)	–	l (m) = 58 ε = 0.5	l (m) = 85 ε = 0.25	–	l (m) = 58 ε = 0.5
	ε					
II	l (m)	55			40	
	ε	0.5			0.67	
III	l (m)	95			48	
	ε	0.33			0.61	
IV	l (m)	152			63	
	ε	0.2			0.52	
V	l (m)	198			80	
	ε	0.13			0.44	
VI	l (m)	–			96	
	ε				0.38	

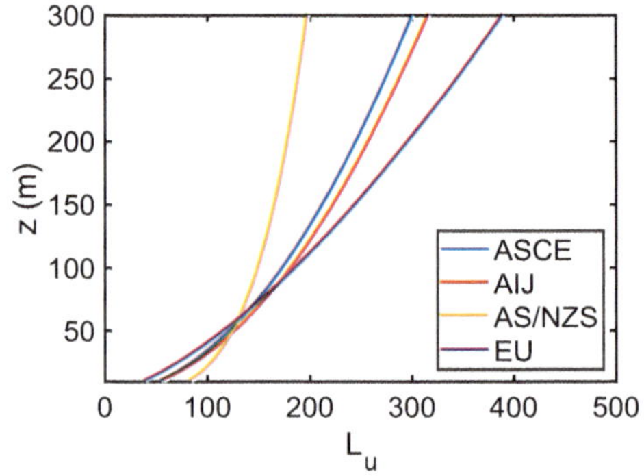

Fig. 5.6 Longitudinal turbulence integral length scale profiles for general urban area stipulated in different codes and standards

$$\frac{L_v^x}{L_u^x} = 0.5\left(\frac{\sigma_v}{\sigma_u}\right)^3 \tag{5.41}$$

$$\frac{L_w^x}{L_u^x} = 0.5\left(\frac{\sigma_w}{\sigma_u}\right)^3 \tag{5.42}$$

where σ_v/σ_u and σ_w/σ_u are from Eqs. (5.9) and (5.10), respectively.

For other components of turbulence integral length scales, the following approximations may be used (Counihan, 1975): $L_u^y \cong 0.3 \sim 0.4L_u^x$ and $L_u^z \cong 0.5 \sim 0.6L_u^x$ for the height range of 10–240 m.

5.1.4 Wind Spectra

As indicated in Sect. 5.1.3, turbulence integral length scales reflect the average eddy sizes. Sometimes we need a detailed information on the turbulent fluctuations and eddies. For example, the resonant response of a flexible structure is triggered by velocity fluctuations with frequencies matching the natural frequencies of the structure (Simiu & Yeo, 2019). To determine this response, it is essential to assess the size of turbulent eddies across different frequencies and how the turbulent fluctuations vary based on their spatial relationships. This information is typically obtained from the wind spectrum and coherence functions.

The basics of the wind spectrum have been introduced in Sect. 1.3. Generally, the models that describe the wind spectrum consider the low-frequency production range and inertial sub-range, where the spectral density S_i ($i = u, v, w$) is a constant and follows the $-$ 5/3 law, respectively. One of the earliest models to describe the wind spectrum is the one proposed by Davenport (1961). However, it is no longer recommended to be used because it does not represent the low-frequency spectral component correctly, and it does not account for the dependence of the spectrum on height. In wind engineering applications, the most widely used form of the wind spectrum is the one proposed by von Kármán (1948):

$$\frac{nS_i}{\sigma_i^2} = \frac{4f_i}{\left(1 + 70.8f_i^2\right)^{5/6}}, \quad i = u \tag{5.43}$$

$$\frac{nS_i}{\sigma_i^2} = \frac{4f_i\left(1 + 755.2f_i^2\right)}{\left(1 + 283.2f_i^2\right)^{11/6}}, \quad i = v, w \tag{5.44}$$

where $f_i = nL_i^x/U$ is the frequency normalized by turbulence integral length scale and wind speed.

Figure 5.7 compares the observed wind spectra at four heights (10, 40, 160, and 320 m) with the von Kármán model. The wind spectra were calculated based on sonic anemometer observations at a tall meteorological tower (Shenzhen Meteorological Tower) in two tropical cyclones (Super Typhoons Mangkhut and Hato). The plot reveals identifiable inertial subranges with a slope of − 2/3, occurring between normalized frequency of 0.2 and 1 for the u and v spectra, and between 0.1 and 0.4 for the w spectrum. Notably, the spectral curves across different heights tend to converge into a single curve, aligning well with the von Kármán framework in both the production and inertial ranges. This behavior is anticipated, as turbulence integral length scale serves as a flexible scaling parameter tailored to fit the estimated scales for specific heights and terrain types while preserving a consistent spectral shape (Teunissen, 1980).

Furthermore, while the inertial subrange exhibits a − 5/3 slope, deviations from this law are noticeable in the high-frequency range, particularly at near-surface levels (e.g., 10 and 40 m). This phenomenon may be attributed to small quasi-independent eddies generated by dynamic near-surface processes. Li et al. (2015) proposed a framework suggesting that the extra energy observed between the inertial subrange and the dissipation range could stem from shear associated with the movement of spray droplets and their evaporation near the surface. The deviations from the − 5/3 law noted in this study may be explained by similar mechanisms as those described by

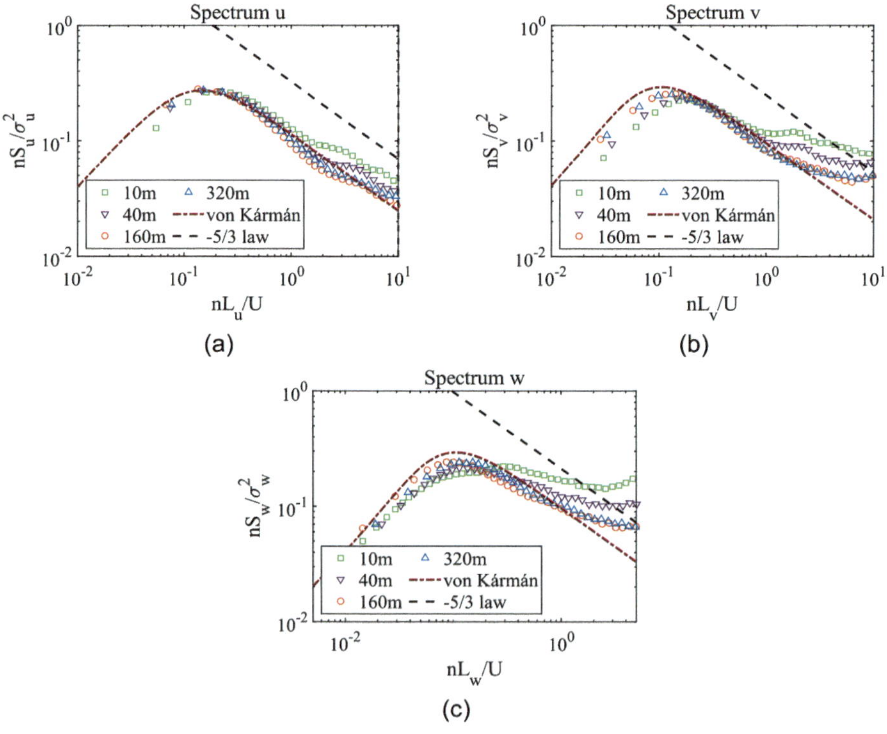

Fig. 5.7 Comparison of observed wind spectra with the von Kármán model (He et al., 2022)

Li et al. (2015). Additionally, it is noted that the energy in the high-frequency range diminishes with increasing height, likely due to the rapid decrease in contributions from near-surface processes as altitude increases (Li et al., 2015).

ESDU (2001) recognized that there are deficiencies of the von Kármán spectrum in describing the wind spectrum at heights near the surface (below about 100 m): (1) L_i^x as derived using an integration of autocorrelation functions tend to be larger than those obtained by fitting the von Kármán spectral equation to measured data. (2) Observed peak values of nS_i/σ_i^2 tend to be smaller than those given by the von Kármán model. To improve the von Kármán spectrum, ESDU (2001) recommended the following spectral model:

$$\frac{nS_i}{\sigma_i^2} = \beta_1 \frac{2.987 f_i/\alpha}{\left[1 + (2\pi f_i/\alpha)^2\right]^{5/6}} + F_1 \beta_2 \frac{1.294 f_i/\alpha}{\left[1 + (\pi f_i/\alpha)^2\right]^{5/6}}, \quad i = u \tag{5.45}$$

$$\begin{aligned}\frac{nS_i}{\sigma_i^2} &= \beta_1 \frac{2.987\left[1 + (8/3)(4\pi f_i/\alpha)^2\right](f_i/\alpha)}{\left[1 + (4\pi f_i/\alpha)^2\right]^{11/6}} \\ &+ F_2 \beta_2 \frac{1.294 f_i/\alpha}{\left[1 + (2\pi f_i/\alpha)^2\right]^{5/6}}, \quad i = v, w\end{aligned} \tag{5.46}$$

$$\alpha = 0.535 + 2.76(0.138 - A)^{0.68} \tag{5.47}$$

$$\beta_1 = 2.357\alpha - 0.761 \tag{5.48}$$

$$\beta_2 = 1 - \beta_1 \tag{5.49}$$

$$F_1 = 1 + 0.455 \exp\left[-0.76\left(\frac{f_i}{\alpha}\right)^{-0.8}\right] \tag{5.50}$$

$$F_2 = 1 + 2.88 \exp\left[-0.218\left(\frac{f_i}{\alpha}\right)^{-0.9}\right] \tag{5.51}$$

where A is given by Eq. (5.36).

Solari and Piccardo (2001) suggested a simple and conservative alternative to the ESDU spectrum:

$$\frac{nS_i}{\sigma_i^2} = \frac{6.868 f_i}{(1 + 10.302 f_i)^{5/3}}, \quad i = u \tag{5.52}$$

$$\frac{nS_i}{\sigma_i^2} = \frac{9.434 f_i}{(1 + 14.151 f_i)^{5/3}}, \quad i = v, w \tag{5.53}$$

In addition to the von Kármán spectrum and its variants, the most widely accepted models for representing the wind spectrum in the ABL are the "blunt" and "point" models. The former is conventionally used for the spectra for the along-wind and cross-wind velocity components, while the latter is frequently adopted for the spectrum for the vertical velocity component. They are given by Eq. (5.54) (blunt) and Eq. (5.55) (point), as follows:

$$\frac{nS_i}{u_*^2} = \frac{Af_z}{(1 + Bf_z)^{5/3}}, \quad i = u, v \tag{5.54}$$

$$\frac{nS_i}{u_*^2} = \frac{Af_z}{1 + Bf_z^{5/3}}, \quad i = w \tag{5.55}$$

where $f_z = nz/U$ is the frequency normalized by height and wind speed, and A and B are coefficients influencing the magnitude and position of the spectral peak. Specifically, the spectral peak occurs at normalized frequency $f_p = n_p z/U = 1.5/B$ (blunt) or $f_p = n_p z/U = 1.27/B^{0.6}$ (point) and corresponds to normalized spectral density $n_p S_p/u_*^2 = 0.33A/B$ (blunt) or $n_p S_p/u_*^2 = 0.51A/B^{0.6}$ (point), where n_p is the peak frequency in Hz. Therefore, an increase in B will shift the spectral peak to a lower frequency, while an increase in A/B (blunt) or $A/B^{0.6}$ (point) will amplify the magnitude of peak spectral density.

Based on the Kansas experiment, Kaimal et al. (1972) established the following form of wind spectra for the neutral surface layer, which has been frequently used in micrometeorological studies:

$$\frac{nS_u}{u_*^2} = \frac{105f_z}{(1 + 33f_z)^{5/3}} \tag{5.56}$$

$$\frac{nS_v}{u_*^2} = \frac{17f_z}{(1 + 9.5f_z)^{5/3}} \tag{5.57}$$

$$\frac{nS_w}{u_*^2} = \frac{2.1f_z}{1 + 5.3f_z^{5/3}} \tag{5.58}$$

Considering that $\int_0^\infty S_u(n)dn = \sigma_u^2$, Simiu and Yeo (2019) modified the Kaimal spectrum such that the area under the spectral curve is approximately $6u_*^2$ (so as to satisfying $(\sigma_u/u_*)^2 = 6$):

$$\frac{nS_u}{u_*^2} = \frac{200f_z}{(1 + 50f_z)^{5/3}} \tag{5.59}$$

Tieleman (1995) recommended the following neutral surface layer spectral models (here shows the blunt model for u and v and point model for w):

$$\frac{nS_u}{u_*^2} = \frac{252.625f_z}{(1+60.62f_z)^{5/3}} \tag{5.60}$$

$$\frac{nS_v}{u_*^2} = \frac{53.78f_z}{(1+20.16f_z)^{5/3}} \tag{5.61}$$

$$\frac{nS_w}{u_*^2} = \frac{2.61f_z}{1+7.23f_z^{5/3}} \tag{5.62}$$

Figure 5.8 compares the observed wind spectra at four heights (10, 40, 160, and 320 m) with the Kaimal model and the Tieleman model. The ensemble averages of the spectral densities are plotted on a log–log scale using the normalization method of Kaimal et al. (1972). The results indicate that the spectral curves at various heights do not converge into a single pattern. Specifically, the spectra at 10 m resemble the Tieleman spectral model, while those at 40 m align with the Kaimal spectral model. However, at heights of 160 and 320 m, both models significantly underestimate the frequency at which the spectral peak occurs. The variation in spectral characteristics from 10 to 320 m suggests that the dominant wavelengths of energy-containing turbulent eddies do not scale effectively with height. This implies that the traditional surface-layer scaling method based on MOST may not be applicable at this location. One reason for this is that 160 and 320 m likely exceed the surface layer where MOST is relevant.

There are also other notable spectral models for the ABL, such as Mann (1994), which has been used in the wind turbine design codes. With respect to the wind codes and standards, the Davenport spectrum is used in CNS and NBCC, the Kaimal spectrum is adopted in ASCE and EU, and the von Kármán spectrum is used in AIJ, AS/NZS, and ISO.

5.1.5 Coherence

The coherence (or normalized cross spectra) function represents the frequency-dependent correlation between two points, which can be defined as:

$$\gamma_{ij}^2(\mathbf{x}_1, \mathbf{x}_2) = \frac{\left|S_{ij}(\mathbf{x}_1, \mathbf{x}_2)\right|^2}{S_i(\mathbf{x}_1)S_j(\mathbf{x}_2)}, \quad i, j = u, v, w \tag{5.63}$$

where $\gamma_{ij}(\mathbf{x}_1, \mathbf{x}_2)$ is the coherence, $S_i(\mathbf{x}_1)$ and $S_j(\mathbf{x}_2)$ are the wind spectra at points $\mathbf{x}_1$ and $\mathbf{x}_2$, respectively, and $S_{ij}(\mathbf{x}_1, \mathbf{x}_2)$ is the cross spectral density between wind velocities at points $\mathbf{x}_1$ and $\mathbf{x}_2$. Sometimes the coherence is defined as γ_{ij}^2 (e.g., Simiu & Yeo, 2019), and γ_{ij} is called the "rooted coherence".

The study of the coherence with $i \neq j$ is not considered in wind engineering applications, such that the notation γ_{ij} is replaced by γ_{ii} in the following for simplicity. The study by Mann (1994) indicates that the effect of blocking of the turbulent eddy

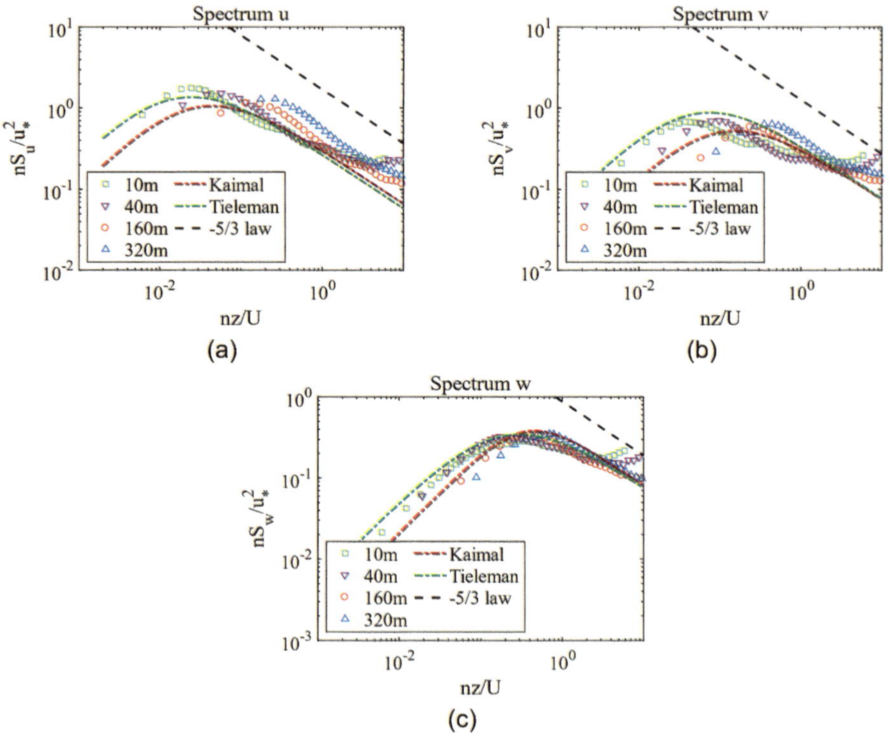

Fig. 5.8 Comparison of observed wind spectra with the Kaimal model and the Tieleman model (He et al., 2022)

by the surface has only a moderate effect on the characteristics of the turbulence. This supports the assumption that the normalised coherence depends on the points $\mathbf{x}_1$ and $\mathbf{x}_2$ mainly through their differences $\mathbf{r} = \mathbf{x}_1 - \mathbf{x}_2 = \left(r_x, r_y, r_z\right)$, therefore $\gamma_{ii}(\mathbf{x}_1, \mathbf{x}_2)$ can be further simplified as $\gamma_{ii}(\mathbf{r})$.

For vertical separations (i.e., $\mathbf{r} = (0, 0, r_z)$), Davenport (1961) proposed one of the earliest and most widely used models of $\gamma_{ii}(r_z)$:

$$\gamma_{ii}(r_z) = \exp\left(-k_i^z \frac{nr_z}{U_r}\right), \quad i = u, v, w \tag{5.64}$$

where k_i^z is an empirical constant known as the decay coefficient, and $U_r = [U(\mathbf{x}_1) + U(\mathbf{x}_2)]/2$.

Later studies argued that the Davenport coherence model is also applicable to longitudinal and lateral separations, i.e., $\gamma_{ii}\left(r_j\right) = \exp\left(-k_i^j nr_j/U_r\right)$, $i = u, v, w$, $j = x, y, z$. Solari and Piccardo (2001) summarized that the average values of the decay coefficients $\left\{k_u^x, k_v^x, k_w^x\right\} = \{3, 3, 0.5\}$, $\left\{k_u^y, k_v^y, k_w^y\right\} = \{10, 6.5, 6.5\}$, and $\left\{k_u^z, k_v^z, k_w^z\right\} = \{10, 6.5, 3\}$.

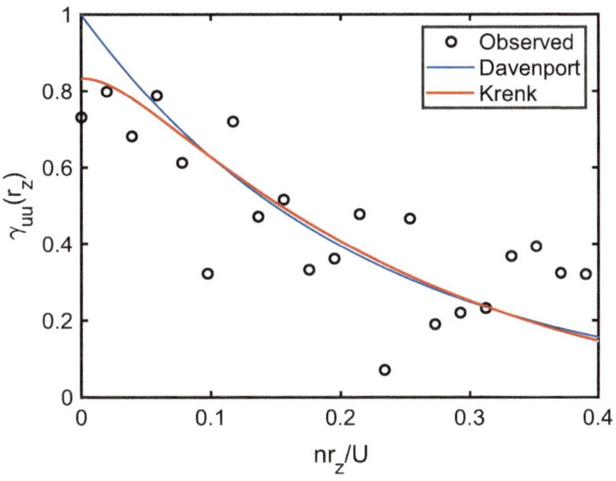

Fig. 5.9 Fitting of observed vertical coherence with the Davenport model and the Krenk model (He et al., 2020)

To better reproduce the coherence as n tends to zero, a modified model was proposed by Krenk for lateral and vertical separations and longitudinal turbulence component (Hansen & Krenk, 1999):

$$\gamma_{uu}(r_y, r_z) = \left[1 - \frac{1}{2}\sqrt{(C_u^y r_y)^2 + (C_u^z r_z)^2}\frac{n_x}{U_r}\right]\exp\left[-\sqrt{(C_u^y r_y)^2 + (C_u^z r_z)^2}\frac{n_x}{U_r}\right] \tag{5.65}$$

$$n_x = \sqrt{n^2 + \left(\frac{U_r}{2\pi L_u^x}\right)^2} \tag{5.66}$$

where C_u^y and C_u^z are empirical constants.

Figure 5.9 shows the vertical coherence between two heights (10 and 40 m), fitted by the model and the Krenk model. The coherence was calculated based on sonic anemometer observations at a tall meteorological tower (Shenzhen Meteorological Tower) in two tropical cyclones (Super Typhoons Mangkhut and Hato). Both Davenport and Krenk decay models showed reasonable agreements with the observed coherence function.

5.2 Turbulence in Non-neutral Conditions

The expressions presented in Sect. 5.1 are typically applicable only under neutral conditions. In non-neutral conditions, atmospheric turbulence exhibits distinct behavior. This section will provide a brief introduction to turbulence in non-neutral conditions, focusing on turbulence intensity and wind spectrum.

As has been introduced in Sect. 2.4.3, the flux-variance similarity can be extended to non-neutral conditions, which provides a way for developing models that describe turbulence intensity under such conditions. A general form of $\sqrt{\beta_i} = \sigma_i/u_*, i = u, v, w$ for non-neutral conditions is given by (Foken et al., 1991; Thomas & Foken, 1995):

$$\sigma_i/u_* = c_{1i}(z/L)^{c_{2i}}, \quad i = u, v, w \tag{5.67}$$

where c_{1i} and c_{2i} are constants. Foken et al. (1991) and Thomas and Foken (1995) suggested that, for slightly unstable conditions ($-0.032 < z/L < 0$), $c_{1u} = 2.7$, $c_{2u} = 0$, $c_{1w} = 1.3$, and $c_{2w} = 0$; for unstable conditions ($z/L < -0.032$), $c_{1u} = 4.15$, $c_{2u} = 0.125$, $c_{1w} = 2.0$, and $c_{2w} = 0.125$. Figure 5.10 shows the variation of σ_u/u_* and σ_w/u_* with stability parameter according to this relationship.

For vertical wind velocity variance, there are generally no substantial differences among the existing parameterizations. The most widely used parameterization, developed by Panofsky and Dutton (1984), is applicable in both neutral and unstable conditions:

$$\sigma_w/u_* = 1.3\left(1 - 2\frac{z}{L}\right)^{1/3} \tag{5.68}$$

Turbulence intensity in non-neutral conditions can then be expressed as:

$$I_i = \frac{\sigma_i}{u_*}\frac{u_*}{U} = \frac{c_{1i}(z/L)^{c_{2i}}k}{\ln[(z/z_0) + \psi_m(z/L)]}, \quad i = u, v, w \tag{5.69}$$

where the stability correction function $\psi_m(\zeta)$ has been given in Sect. 4.2.1.

The wind spectrum also looks different in non-neutral conditions. According to the spectrum similarity theory (Sect. 2.4.4), the power spectral density for high

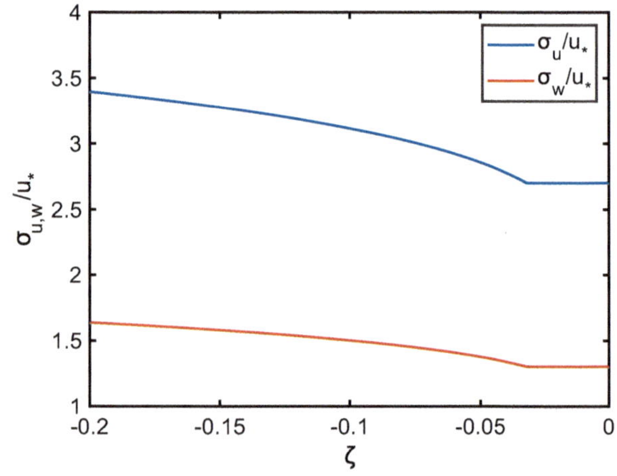

Fig. 5.10 Variation of σ_u/u_* and σ_w/u_* with stability parameter

frequencies ($f_z > 4$) can be modelled as (Kaimal et al., 1972):

$$\frac{nS_i}{u_*^2} = \frac{\alpha_u}{(2\pi k)^{2/3}}\varphi_\varepsilon^{2/3} f_z^{-2/3}, \quad i = u \tag{5.70}$$

$$\frac{nS_i}{u_*^2} = 0.4\varphi_\varepsilon^{2/3} f_z^{-2/3}, \quad i = v, w \tag{5.71}$$

where $f_z = nz/U$, φ_ε is the non-dimensional dissipation rate for turbulent kinetic energy. The Kansas experiment gives the following expressions for φ_ε (Kaimal et al., 1972) as a function of z/L:

$$\text{For } -2 < z/L < 0: \quad \varphi_\varepsilon^{\frac{2}{3}}\left(\frac{z}{L}\right) = 1 + 0.5\left|\frac{z}{L}\right|^{2/3} \tag{5.72}$$

$$\text{For } 0 < z/L < 2: \quad \varphi_\varepsilon^{2/3}\left(\frac{z}{L}\right) = 1 + 2.5\left|\frac{z}{L}\right|^{3/5} \tag{5.73}$$

For low frequencies ($f_z < 4$), the shapes of the spectral density are similar to those given by Eqs. (5.56)–(5.58).

5.3 Turbulence in Storms

5.3.1 Turbulence in Tropical Cyclones

Turbulence characteristics in tropical cyclones (TCs) are generally considered to differ from those in other synoptic-scale wind events (e.g., extratropical cyclones and monsoons). Despite extensive research, our understanding of the similarities and differences in turbulence characteristics between TCs and synoptic winds remains limited. While some comparative studies based on near-surface observations have been conducted, their findings are often fragmented.

Some research indicates that turbulence intensity (and consequently gust factors) is higher in TCs compared to synoptic winds. For instance, Choi (1978) noted that longitudinal turbulence intensity in tropical cyclone winds can be approximately 50% greater than that in synoptic winds. Other studies, such as Schroeder and Smith (2003), Sharma and Richards (1999) and Yu and Chowdhury (2009), observed a more modest increase of about 10%. Conversely, other studies, such as those by Vickery and Skerlj (2005) and Shu et al. (2015), suggest that turbulence intensities and gust factors in TCs and synoptic winds are nearly equivalent.

The increased turbulence intensity observed in TCs may be linked to the convective "squall-like" turbulence that occurs in TCs. This type of turbulence can be significantly more intense than the mechanical turbulence typically found near the surface and is associated with the movement of rain cloud bands.

Figure 5.11 illustrates the vertical distributions of longitudinal turbulence intensity during typhoons and synoptic winds (monsoons) in both suburban and open terrain conditions. The turbulence intensity was calculated based on sonic anemometer observations at a tall meteorological tower (Shenzhen Meteorological Tower). Generally, the turbulence intensity profiles can be effectively described by the power law. In suburban areas (Fig. 5.11a), the longitudinal turbulence intensity I_u at 10 and 40 m during typhoons and monsoons aligns closely with values recommended by ASCE, but is typically higher than those suggested by GB, AIJ, NBCC, EU, and ISO. At 160 and 320 m, I_u for typhoons approaches the ISO recommendations, while for monsoons, it is similar to values from GB, AIJ, NBCC, and AS/NZS. Notably, the coefficient c for both typhoons and monsoons is consistent with ASCE's recommendation at 0.3, although the exponent d is slightly larger for typhoons (0.22) and monsoons (0.27) compared to ASCE's 0.167. In open terrain (Fig. 5.11b), the values of c are higher—0.26 for typhoons and 0.28 for monsoons—compared to the typical standard of around 0.20. This discrepancy may arise because the mean roughness length for the analyzed open terrain is greater than the typical 0.03 m. Additionally, turbulence intensity in suburban areas is generally greater than in open terrain for both typhoons and monsoons.

Comparing turbulence intensity profiles, I_u at near-surface levels (10 and 40 m) is nearly identical for both typhoons and monsoons. However, at higher altitudes (160 and 320 m), I_u for typhoons is 15–30% greater than for monsoons, resulting in a flatter profile with a lower d value for typhoons. The observed differences in turbulence intensity between typhoon and monsoon winds may stem from their respective boundary layer characteristics. In near-neutral conditions (where buoyancy effects are minimal), turbulence in the surface layer (10 and 40 m) is primarily generated by mechanical wind shear. Therefore, turbulence intensity increases with rougher upwind terrain. However, at higher elevations (160 and 320 m), within the outer layer (Ekman layer), σ_u/u_* varies with the nondimensional height z/h. Since

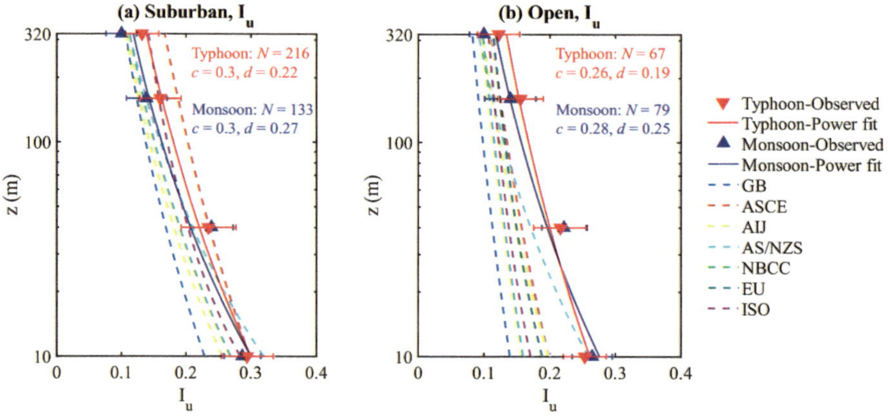

Fig. 5.11 Longitudinal turbulence intensity profiles in typhoons and monsoons (He et al., 2021)

the ABL height is generally higher during typhoons than monsoons, this results in a lower z/h ratio, leading to higher turbulence intensity in typhoons. Additionally, convective processes in the typhoon boundary layer may further enhance turbulence intensity at these heights compared to the monsoon boundary layer.

Figure 5.12 illustrates the vertical distributions of turbulence integral length scales during typhoons and monsoons in both suburban and open terrain. The data indicate that the vertical variation of these scales can be effectively described by a power law. In suburban areas (Fig. 5.12a), the turbulence integral length scale L_u for typhoons falls between the values recommended by ASCE and AS/NZS, while L_u for monsoons is slightly below the AS/NZS guideline. Specifically, for typhoons, $l = 87$ m aligns closely with AS/NZS (85 m) but is lower than ASCE's recommendation (95 m). The value of $\epsilon = 0.32$ is higher than the AS/NZS suggestion of 0.25, yet comparable to ASCE's value of 0.33. In open terrain (Fig. 5.12b), L_u for typhoons is similar to ASCE's recommendations, while L_u for monsoons matches the AS/NZS figures. Notably, the turbulence integral length scale is lower for typhoon winds in suburban terrain compared to open terrain, whereas for monsoon winds, L_u is greater in suburban areas than in open ones.

When comparing L_u between typhoons and monsoons, it is evident that L_u is significantly larger during typhoons at all four height levels (10, 40, 160, and 320 m). In open terrain, typhoon values are approximately 45–80% greater than those for monsoons. In suburban terrain, L_u for typhoons is about 20–45% larger than for monsoons. This higher turbulence integral length scale in typhoons is likely linked to the greater ABL height observed during these events, as L_u is considered proportional to the ABL height.

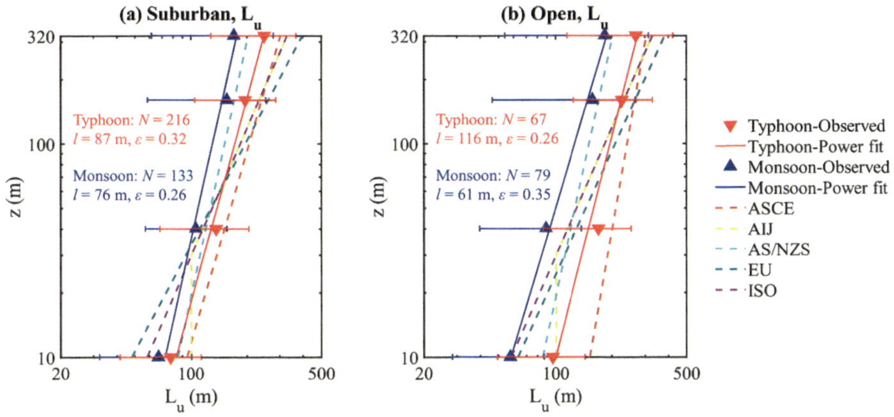

Fig. 5.12 Longitudinal integral length scale profiles in typhoons and monsoons (He et al., 2021)

5.3.2 Turbulence in Thunderstorms

The turbulence characteristics associated with downburst winds in thunderstorms are even less well understood than those in TCs. Figure 5.13 depicts the time-varying mean wind speed at 320-m height in four thunderstorm events. The four events under consideration occur on four days in the summer of 2019. The high wind gust occurred at about the middle of each time period, and data around 20 min before and after the high wind gust were included in the statistical study of the wind associated with the thunderstorms.

For event (a) shown in Fig. 5.13a, the time-varying mean wind speed dramatically increases from 4 to 19.1 m/s, followed by a plateau with constant high wind speeds of roughly 13–16 m/s. During the plateau, two consecutive peaks can be identified, one of 15.4 m/s, and the other of 15 m/s. After that, the time-varying mean wind speed starts to ramp down, and quickly drops from 12.5 to 1.6 m/s. With regard to event (d) shown in Fig. 5.13d, a drastic increase of time-varying mean wind speed from 5 to 23.8 m/s can be observed. After the first peak, the wind speed slightly decreases before rapidly increasing again, and reaches the second peak of 27.2 m/s. Then the wind speed drops slowly with time, from 18.9 to 7.5 m/s. For the other two events, similar ramp-up and ramp-down behaviors can be found.

Figure 5.14 plots the fluctuating wind speed at 320-m height in these events. Significant variation in the fluctuating wind component can be observed. Figures 5.13

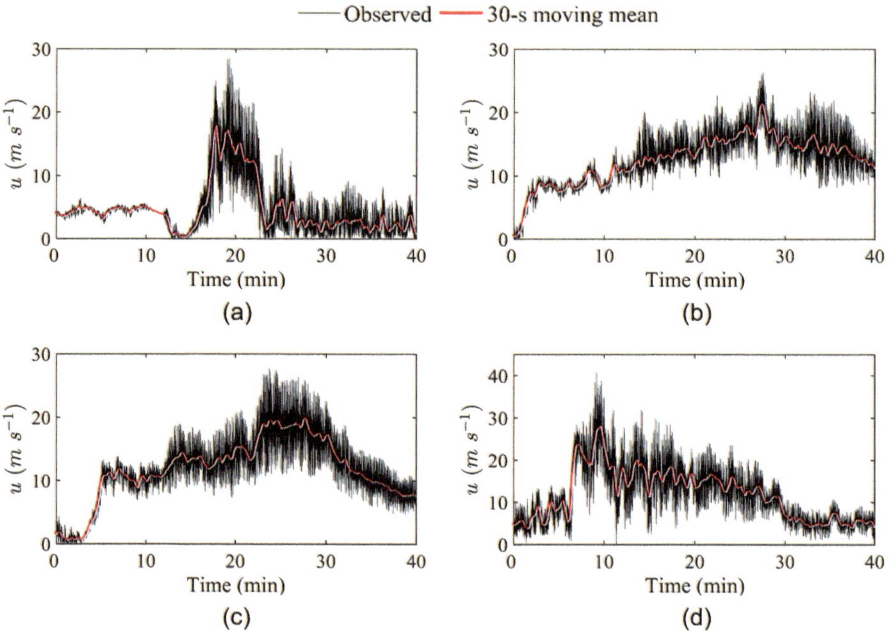

Fig. 5.13 Time-varying mean wind speed for four thunderstorm events

and 5.14 together suggest that strong nonstationarity occurs in not only the mean wind speed but also the fluctuating or turbulent component of wind velocity during thunderstorms. Therefore, both the mean wind and turbulence in thunderstorms, when observed at a fixed point, is highly nonstationary. It is necessary to use the nonstationary wind model so as to estimate the structural loads and responses during thunderstorm wind excitations.

The reduced turbulence fluctuation and its probability distribution are given in Figs. 5.15 and 5.16 (definition and calculation of reduced turbulence fluctuation can be found in He et al. (2021)). The black line in Fig. 5.16 represents the Gaussian distribution. Generally, the Gaussian distribution provides a good fit to the probability distribution of the reduced turbulent fluctuation.

Based on the reduced turbulent fluctuation, the power spectral density is calculated and shown in Fig. 5.17. The − 5/3 slope in the power spectral density curves can be clearly identified in the frequency range between 0.04 and 1 Hz, which is in accordance with Kolmogorov's theory of local isotropic turbulence in the inertial subrange.

To have a closer look on the thunderstorm turbulence intensity profiles, Fig. 5.18 portrays the vertical profiles of time-varying mean turbulence intensity in the thunderstorm event (a). Similar to observations by Zhang et al. (2019) and Canepa et al. (2020), most of the observed turbulence intensity profiles display a zigzag trend with height and a considerable degree of scatter in values. At the beginning of the ramp-up stage, the turbulence intensity profile exhibits an off-mean spike (blue lines) with noticeably large turbulence intensity up to 0.4 at 160 m. As the wind speed continues

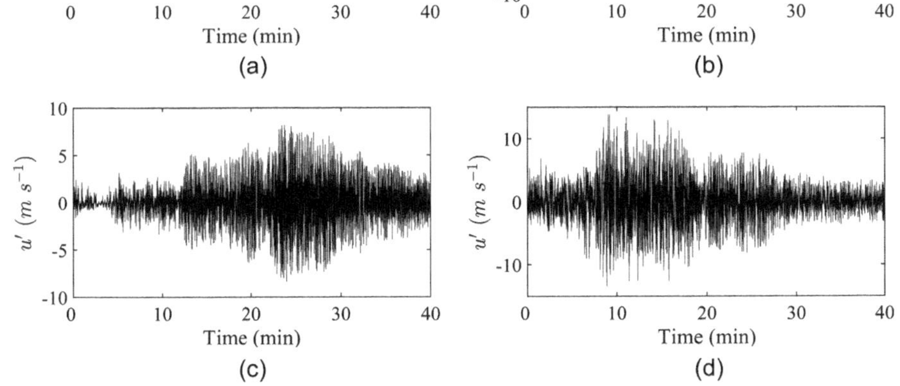

Fig. 5.14 Fluctuating wind speed for four thunderstorm events

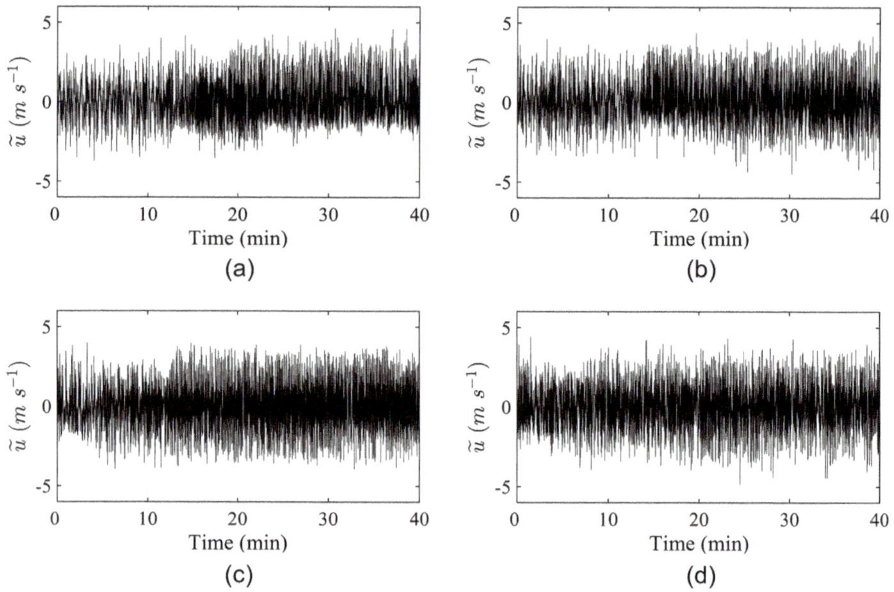

Fig. 5.15 Reduced turbulence fluctuation for four thunderstorm events

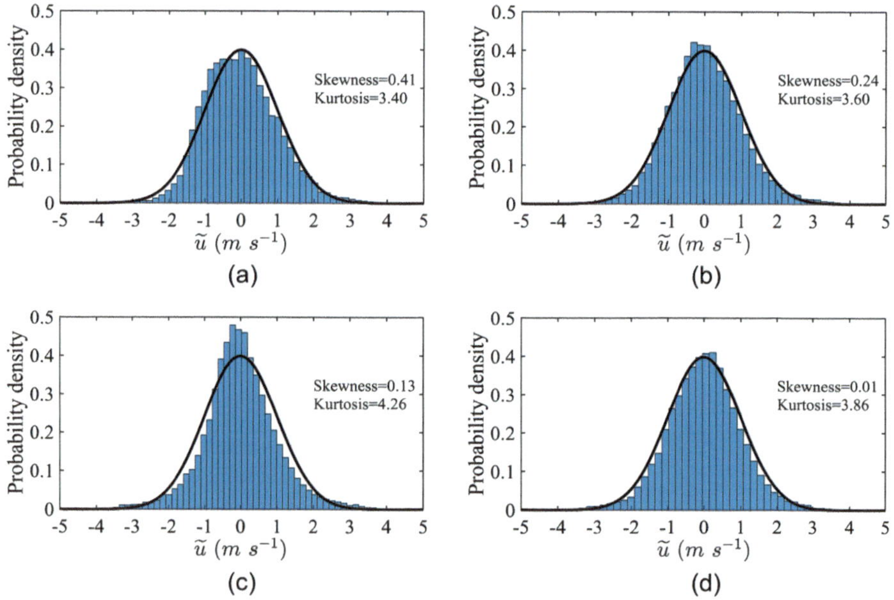

Fig. 5.16 Histograms of reduced turbulence fluctuation for four thunderstorm events

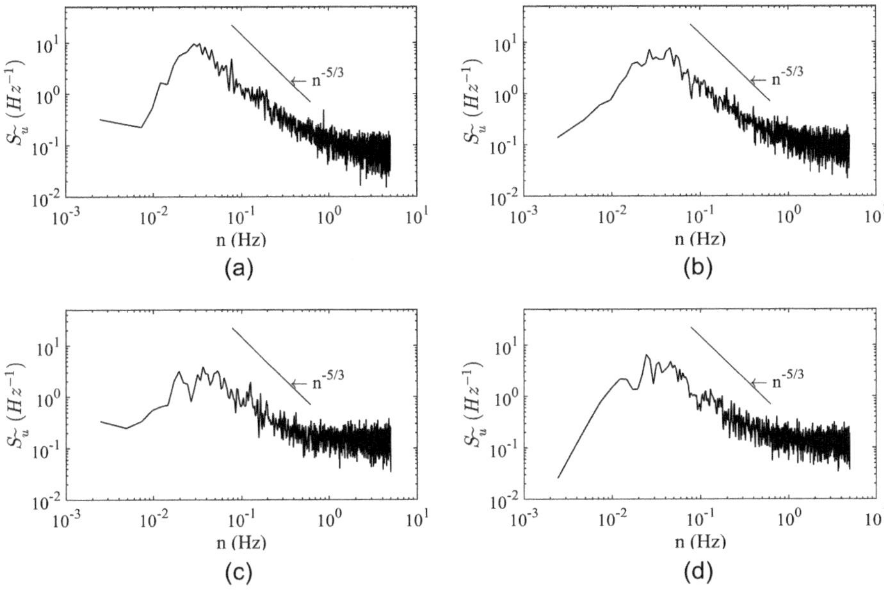

Fig. 5.17 Power spectral density of reduced turbulence fluctuation for four thunderstorm events

to ramp up, the turbulence intensity at 160 m declines rapidly. During the plateau period, there appears to be a nose-shaped form of turbulence intensity profile with the maximum at 40 m. The mean turbulence intensity at 40 m (0.31) is fairly larger than 10 m (0.26), while that at 160 and 320 m is almost the same (0.21). At the ramp-down phase, the turbulence intensity tends to decrease with increasing height, from 0.27 at 10 m to 0.19 at 320 m.

Figure 5.19 shows the vertical profiles of time-varying mean turbulence intensity in the thunderstorm event (d) on 10 June 2019. As the wind speed ramps up between 370 and 400 s, the turbulence intensity profile is generally uniform over height. During the plateau period, a nose-liked shape of turbulence intensity profile can be observed. The mean turbulence intensity increases from 0.23 at 10 m to 0.28 at 40 m, followed by a decrease to 0.22 at 160 m and 0.19 at 320 m. At the ramp-down stage, the turbulence intensity exhibits a somewhat zigzag profile over height, with turbulence intensity of 0.22 at 10 m and 0.15 at 320 m.

During the strongest winds in thunderstorms, both Figs. 5.18 and 5.19 suggests a nose-liked shape of turbulence intensity profiles. Turbulence intensity increases with increasing heights within the lowest 40 m, and decreases aloft. The nose-shaped turbulence intensity profiles in thunderstorms are quite different from TCs and monsoons where turbulence intensity monotonously decreases with increasing height. The values of turbulence intensity in thunderstorms are also significantly larger than those suggested by codes and standards for open exposure. A possible reason is that, the short timescales of thunderstorm events may not provide sufficient time for the wind and turbulence profiles to develop and reach an equilibrium state

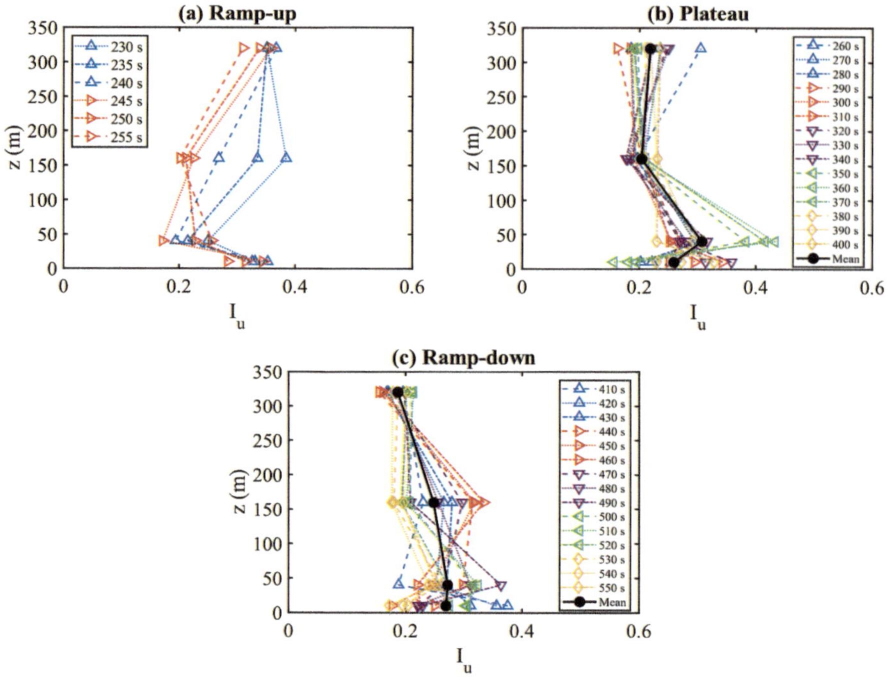

Fig. 5.18 Vertical profiles of time-varying mean longitudinal turbulence intensity in the thunderstorm event (a) (He et al., 2021)

with the underlying surface. Another plausible factor is that, the strong downburst in thunderstorms can significantly enhance the convection contribution to the turbulence, therefore the wind and turbulence structures of thunderstorms show a limited connection with terrain exposure but largely depend on their internal convective structure.

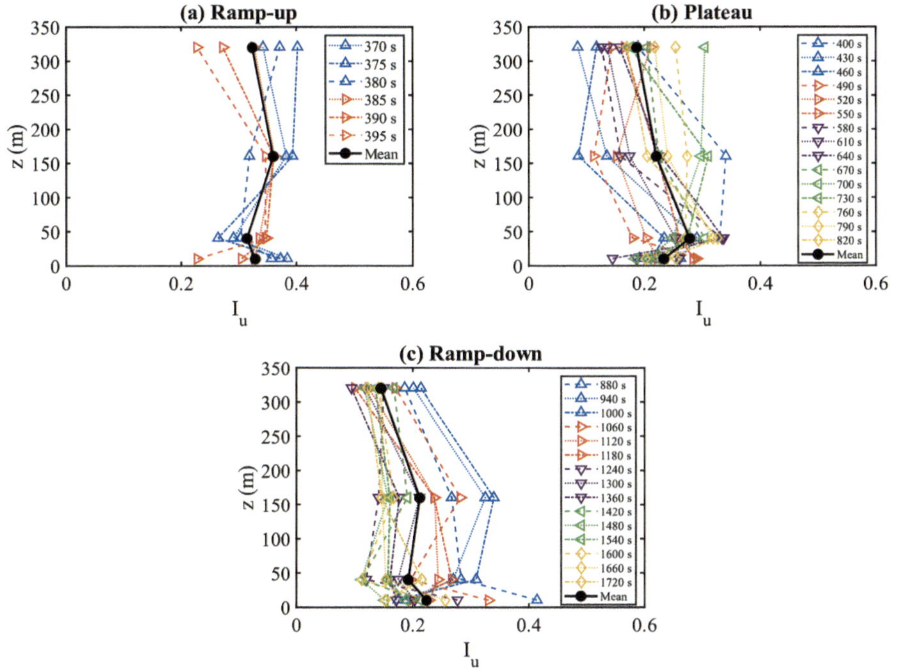

Fig. 5.19 Vertical profiles of time-varying mean longitudinal turbulence intensity in the thunderstorm event (d) (He et al., 2021)

References

Canepa, F., Burlando, M., & Solari, G. (2020). Vertical profile characteristics of thunderstorm outflows. *Journal of Wind Engineering & Industrial Aerodynamics, 206*, 104332. https://doi.org/10.1016/j.jweia.2020.104332

Cao, S., Tamura, Y., Kikuchi, N., Saito, M., Nakayama, I., & Matsuzaki, Y. (2009). Wind characteristics of a strong typhoon. *Journal of Wind Engineering & Industrial Aerodynamics, 97*, 11–21. https://doi.org/10.1016/j.jweia.2008.10.002

Choi, E. C. C. (1978). Characteristics of typhoons over the South China Sea. *Journal of Wind Engineering & Industrial Aerodynamics, 3*, 353–365.

Counihan, J. (1975). Adiabatic atmospheric boundary layers: A review and analysis of data from the period 1880–1972. *Atmospheric Environment (1967), 9*, 871–905. https://doi.org/10.1016/0004-6981(75)90088-8

Davenport, A. G. (1961). A statistical approach to the treatment of wind loading on tall masts and suspension bridges.

Durst, C. S. (1960). Wind speeds over short periods of time. *The Meteorological Magazine, 89*, 181–187.

ESDU. (2001). *Characteristics of atmospheric turbulence near the ground. Part II: Single point data for strong winds (neutral atmosphere). Data item 85020*. Engineering Sciences Data Unit.

ESDU. (2012). *Strong winds in the atmospheric boundary layer. Part 2: Discrete gust speeds. Data item 83045*. Engineering Sciences Data Unit.

Foken, T., Skeib, G., & Richter, S. H. (1991). Dependence of the integral turbulence characteristics on the stability of stratification and their use for Doppler-Sodar measurements. *Meteorologische Zeitschrift, 41*, 311–315.

Hansen, S. O., & Krenk, S. (1999). Dynamic along-wind response of simple structures. *Journal of Wind Engineering & Industrial Aerodynamics, 82*, 147–171. https://doi.org/10.1016/S0167-6105(98)00215-3

He, J. Y., Chan, P. W., Li, Q. S., Li, L., Zhang, L., & Yang, H. L. (2022). Observations of wind and turbulence structures of Super Typhoons Hato and Mangkhut over land from a 356 m high meteorological tower. *Atmospheric Research, 265*, 105910. https://doi.org/10.1016/j.atmosres.2021.105910

He, J. Y., He, Y. C., Li, Q. S., Chan, P. W., Zhang, L., Yang, H. L., & Li, L. (2020). Observational study of wind characteristics, wind speed and turbulence profiles during Super Typhoon Mangkhut. *Journal of Wind Engineering & Industrial Aerodynamics, 206*, 104362. https://doi.org/10.1016/j.jweia.2020.104362

He, J. Y., Li, Q. S., Chan, P. W., Li, L., Lu, C., Zhang, L., & Yang, H. L. (2021). Characteristics and vertical profiles of mean wind and turbulence for typhoon, monsoon, and thunderstorm winds. *Journal of Structural Engineering, 147*, 04021188. https://doi.org/10.1061/(asce)st.1943-541x.0003156

Högström, U., Hunt, J. C. R., & Smedman, A. S. (2002). Theory and measurements for turbulence spectra and variances in the atmospheric neutral surface layer. *Boundary-Layer Meteorology, 103*, 101–124. https://doi.org/10.1023/A:1014579828712

Holmes, J. D. (2015). *Wind loading of structures* (3rd ed.). CRC Press.

Ishizaki, H. (1983). Wind profiles, turbulence intensities and gust factors for design in typhoon-prone regions. *Journal of Wind Engineering & Industrial Aerodynamics, 13*, 55–66.

Kaimal, J. C., Wyngaard, J. C., Izumi, Y., & Cote, O. R. (1972). Spectra characteristics of surface-layer turbulence. *Quarterly Journal of the Royal Meteorological Society, 98*, 563–589.

Kareem, A., & Zhao, J. (1994). Analysis of non-Gaussian surge response of tension leg platforms under wind loads. *Journal of Offshore Mechanics and Arctic Engineering, 116*, 137–144. https://doi.org/10.1115/1.2920142

Kwon, D. K., & Kareem, A. (2013). Comparative study of major international wind codes and standards for wind effects on tall buildings. *Engineering Structures, 51*, 23–35. https://doi.org/10.1016/j.engstruct.2013.01.008

Li, L., Kareem, A., Hunt, J., Xiao, Y., Zhou, C., & Song, L. (2015). Turbulence spectra for boundary-layer winds in tropical cyclones: A conceptual framework and field measurements at coastlines. *Boundary-Layer Meteorology, 154*, 243–263. https://doi.org/10.1007/s10546-014-9974-7

Mann, J. (1994). The spatial structure of neutral atmospheric surface-layer turbulence. *Journal of Fluid Mechanics, 273*, 141–168. https://doi.org/10.1017/S0022112094001886

Panofsky, H. A., & Dutton, J. A. (1984). *Atmospheric turbulence, models and methods for engineering applications* (1st ed.). Wiley.

Schroeder, J. L., & Smith, D. A. (2003). Hurricane Bonnie wind flow characteristics as determined from WEMITE. *Journal of Wind Engineering & Industrial Aerodynamics, 91*, 767–789. https://doi.org/10.1016/S0167-6105(02)00475-0

Sharma, R. N., & Richards, P. J. (1999). A re-examination of the characteristics of tropical cyclone winds. *Journal of Wind Engineering & Industrial Aerodynamics, 83*, 21–33. https://doi.org/10.1016/S0167-6105(99)00058-6

Shu, Z. R., Li, Q. S., He, Y. C., & Chan, P. W. (2015). Gust factors for tropical cyclone, monsoon and thunderstorm winds. *Journal of Wind Engineering & Industrial Aerodynamics, 142*, 1–14. https://doi.org/10.1016/j.jweia.2015.02.003

Simiu, E., Potra, F. A., & Nandi, T. N. (2019). Determining longitudinal integral turbulence scales in the near-neutral atmospheric surface layer. *Boundary-Layer Meteorology, 170*, 349–355. https://doi.org/10.1007/s10546-018-0400-4

Simiu, E., & Yeo, D. H. (2019). *Wind effects on structures: Modern structural design for wind* (4th ed.). Wiley.

Solari, G., & Piccardo, G. (2001). Probabilistic 3-D turbulence modeling for gust buffeting of structures. *Probabilistic Engineering Mechanics, 16*, 73–86. https://doi.org/10.1016/S0266-8920(00)00010-2

Teunissen, H. W. (1980). Structure of mean winds and turbulence in the planetary boundary layer over rural terrain. *Boundary-Layer Meteorology, 19*, 187–221. https://doi.org/10.1007/BF00117220

Thomas, C., & Foken, T. (1995). Re-evaluation of integral turbulence characteristics and their parameterisations (pp. 2–5).

Tieleman, H. W. (1995). Universality of velocity spectra. *Journal of Wind Engineering & Industrial Aerodynamics, 56*, 55–69. https://doi.org/10.1016/0167-6105(94)00011-2

Vickery, P. J., & Skerlj, P. F. (2005). Hurricane gust factors revisited. *Journal of Structural Engineering, 131*, 825–832. https://doi.org/10.1061/(ASCE)0733-9445(2005)131:5(825)

von Kármán, T. (1948). Progress in the statistical theory of turbulence. *Proceedings of the National Academy of Sciences, 34*, 530–539. https://doi.org/10.1073/pnas.34.11.530

Yu, B., & Chowdhury, A. G. (2009). Gust factors and turbulence intensities for the tropical cyclone environment. *Journal of Applied Meteorology and Climatology, 48*, 534–552. https://doi.org/10.1175/2008JAMC1906.1

Zhang, S., Yang, Q., Solari, G., Li, B., & Huang, G. (2019). Characteristics of thunderstorm outflows in Beijing urban area. *Journal of Wind Engineering & Industrial Aerodynamics, 195*, 104011. https://doi.org/10.1016/j.jweia.2019.104011

Chapter 6
Winds Over Complex Terrain

The models presented in Chaps. 4 and 5 outline the mean wind and turbulence in the atmospheric boundary layer under ideal conditions, i.e., horizontally uniform surfaces without obstacles and steady-state scenarios. However, these ideal conditions are rarely encountered in reality, as the Earth's surface is never completely flat and surface characteristics vary.

This chapter describes and explores the impact of heterogeneous surfaces on wind and turbulence in the atmospheric boundary layer. Topics mainly include roughness change (Sect. 6.1), topographic effect (Sect. 6.2), urban surface (Sect. 6.3), and sea surface (Sect. 6.4). Given the significant influence of these factors, they need to be carefully considered in the measurements and modeling of the atmospheric boundary layer.

6.1 Roughness Change and Internal Boundary Layer

Chapters 4 and 5 have focused on the assumption of a homogeneous surface; however, most landscapes feature heterogeneous surfaces with varying characteristics over distances of about 100 m. Figure 6.1 illustrates the complex flow systems that develop above such inhomogeneous terrain. Each surface generates a distinct wind profile based on its roughness or a temperature profile based on its surface temperature. The horizontal wind field shifts these varying profiles downstream, creating a discontinuity layer known as the internal boundary layer (IBL). This means that neighboring areas can influence the exchange of energy and matter above different downstream regions.

IBLs arise from variations in surface roughness and differences in temperature, moisture, and other factors. They can exist under neutral, stable, and unstable atmospheric conditions, with vertical exchanges becoming more dominant during free convection. These IBLs may comprise several layers of varying thickness. The layer

J. He and Q.-S. Li, *Winds in the Atmospheric Boundary Layer*,
https://doi.org/10.1007/978-981-95-1517-2_6

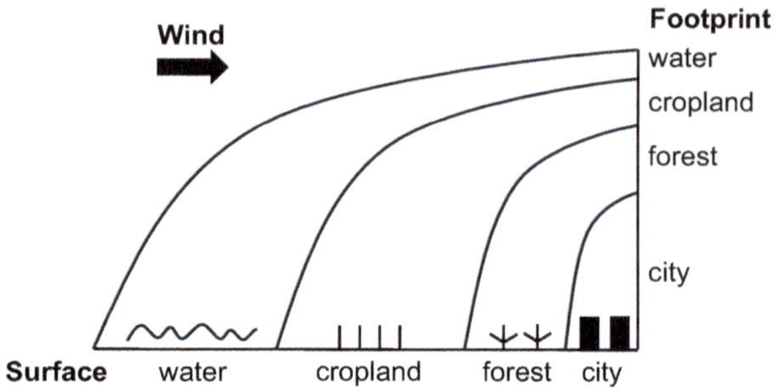

Fig. 6.1 Generation of internal boundary layer above heterogeneous surfaces

beneath the discontinuity is called the new equilibrium layer, while the layer above is influenced by the upwind surface conditions.

The development of mechanical IBLs is affected by roughness lengths and friction velocities on either side of a sudden change in surface roughness. The height of the IBL is influenced by the fetch, or distance to the roughness change; larger fetches generally reduce the differences across the IBL. Figure 6.2 shows the illustrative wind profiles for internal boundary layer developing over rough-to-smooth change and smooth-to-rough change in neutral atmospheric stability conditions.

Several methods exist to determine the height of the IBL from mean wind profiles. The simplest approach involves extrapolating wind profiles from both sides of the IBL, although this can sometimes place the point of intersection outside the IBL. The height of the IBL can also be determined by analyzing changes in friction velocity. Modelers often use the friction velocity approach for this purpose. It is important to note that the IBL defined through the friction velocity method tends to be taller than those calculated from wind profiles.

Elliott (1958) provided a classical model for estimating the height of the IBL δ:

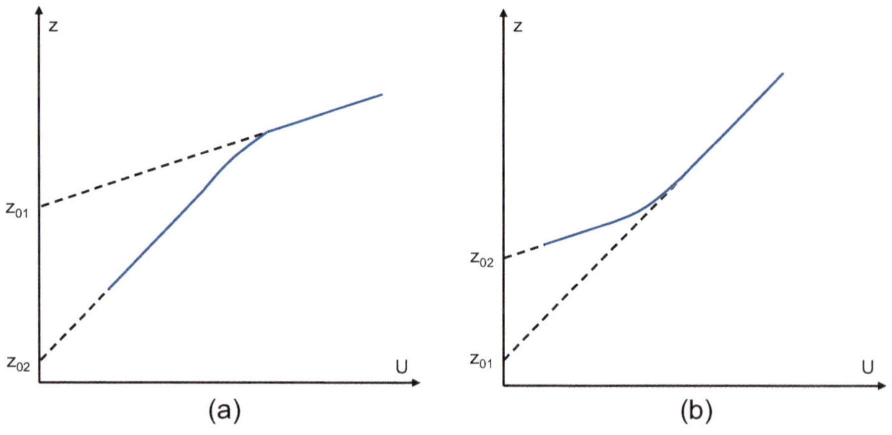

Fig. 6.2 Wind profiles for internal boundary layer developing over **a** rough-to-smooth change and **b** smooth-to-rough change in neutral atmospheric stability conditions

$$\delta = z_{02}\left(\frac{z_{01}}{z_{02}}\right)^{1/[1-(u_{*2}/u_{*1})]} \tag{6.1}$$

where z_{01} and z_{02} are roughness lengths before and after the roughness change, respectively, and u_{*1} and u_{*2} are friction velocities before and after the roughness change, respectively.

Another parameterization of the height of the IBL is (Logan & Fichtl, 1975):

$$\frac{u_{*2}}{u_{*1}} = 1 - \frac{\ln(z_{01}/z_{02})}{\ln(\delta/z_{01})} \tag{6.2}$$

A simple model of IBL height given by Wood (1982) is:

$$\delta = 0.28 z_{0r}\left(\frac{x}{z_{0r}}\right)^{0.8} \tag{6.3}$$

where z_{0r} is the larger of z_{01} and z_{02}, and x is the downwind distance after the roughness change.

Figure 6.3 shows wind profiles after a sea-land roughness change, which was observed by radiosondes over the Dongsha Island. The wind profile above 100 m over the offshore island resembles that of the deep open sea. However, below 100 m, the wind profile, while logarithmic, exhibits a steeper vertical gradient in linear-logarithmic space compared to the higher altitudes. This difference arises from the transition between sea and land at the island's coastline. The IBL below 100 m is mainly affected by the rougher rural land surface of the island, whereas the outer layer above 100 m reflects the smoother sea surface surrounding the island.

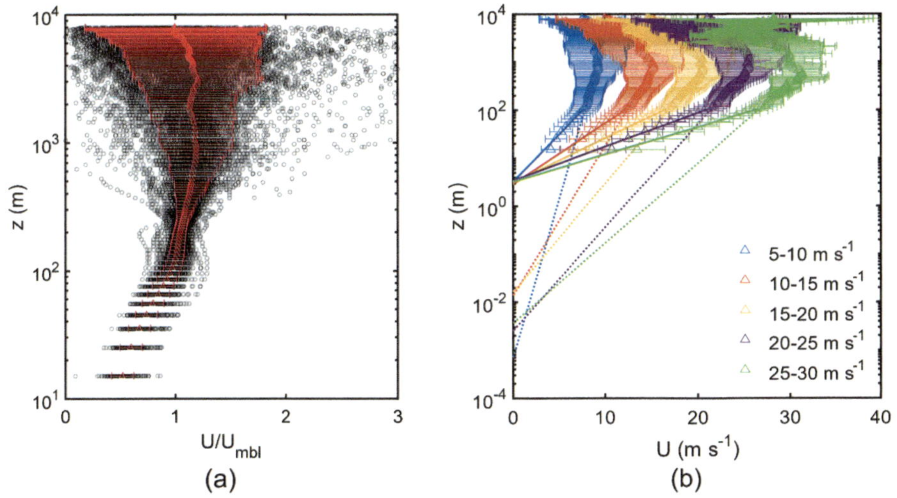

Fig. 6.3 Observation of wind profiles after a sea-land roughness change

6.2 Topographic Effects

Natural and man-made topography, such as escarpments, ridges, and hills, can significantly enhance wind speeds. Research in the 1970s and 1980s improved predictions of wind speeds over shallow topography, while the effects of turbulence and steep topography remain less well understood, which is crucial for structural design.

As wind approaches a shallow topographic feature, its speed initially decreases slightly but increases as it ascends, peaking at or just before the crest. Wind speed then gradually decreases after the crest. Over ridges with downwind slopes, this adjustment occurs more rapidly than over escarpments. For steeper topographic features, flow separation can occur due to increasing pressure gradients, potentially at the upwind slope, just beyond the crest, or along the downwind slope. In steep cases (greater than about 0.3), the upwind separation bubble presents an effective slope, often used to limit speed-up effects in engineering standards (Holmes, 2015).

Speed-up effects are greatest near the surface and diminish with height, leading to mean velocity profiles that may be constant or peaked near the crest. Current codes and standards mainly focus on two-dimensional (2-D) features, which are effective for long ridges and escarpments. However, three-dimensional (3-D) effects, such as airflow around hills, can reduce speeds over the top. For structural design, it is common to ignore these effects and calculate wind loads based solely on the upwind and downwind slopes (Holmes, 2015).

Figure 6.4 shows the simulation of wind speed over a topographic feature (Cheung Chau Island in Hong Kong) using CFD. The spatial variation in wind speed is influenced by altitude. The wind speed increases with higher elevations due to the topographic speed-up effect. The peak wind speed occurs at the hilltop, reaching approximately 4.5 m/s.

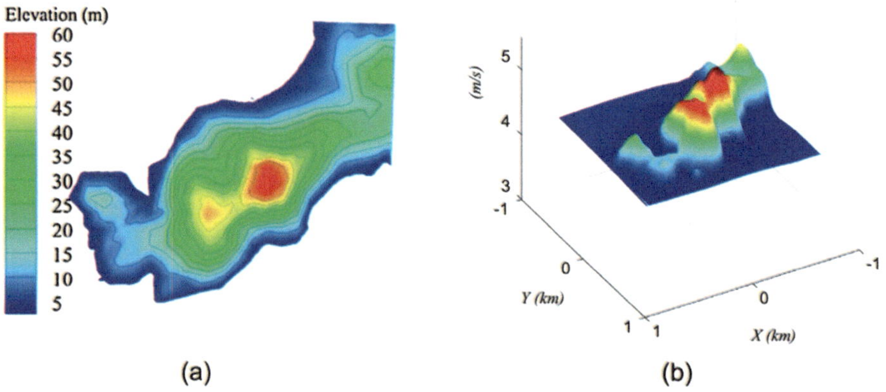

Fig. 6.4 Winds over a topographic feature, **a** surface elevation, and **b** spatial distribution of wind speed from CFD simulation (Yan & Li, 2016)

In structural design codes and standards, the topographic factor (or topographic multiplier) is usually used to quantify the speed-up effect. It is defined as the ratio between the wind speed above the topographic feature and that above the flat ground upwind. The ASCE standard gives the following expression of the topographic factor K_z for 2-D or 3-D isolated hills as well as on 2-D ridges and escarpments:

$$K_z = 1 + K_1 K_2 K_3 \tag{6.4}$$

$$K_1 = \frac{aH}{L_h} \tag{6.5}$$

$$K_2 = 1 - \frac{x}{|\mu L_h|} \tag{6.6}$$

$$K_3 = \exp\left(-\frac{\gamma z}{L_h}\right) \tag{6.7}$$

where H is the height of the topographic feature, L_h is the half of the upwind length of the feature, and x is the horizontal distance from the crest. The above expressions for K_1, K_2, and K_3 are valid for $H/L_h < 0.5$. For 2-D escarpment, $\gamma = 2.5$, $\mu = 1.5$ for upwind of crest and 4.0 for downwind of crest, and $a = 0.75$ for Exposure B, 0.85 for Exposure C, and 0.95 for Exposure D.

6.3 Winds Over Cities

Studies of the urban boundary layer (UBL) have gained attention due to the rise of megacities, where over half of the global population resides. Cities are significant pollution sources and more vulnerable to climate change, prompting investigations into UBL dynamics, such as wind and turbulence profiles as well as urban heat islands. Urban areas feature rough surfaces and limited moisture, leading to higher turbulence and greater sensible heat fluxes. This results in elevated UBL temperatures, contributing to the urban heat island effect, which can be intensified by human energy production. Urban heat islands generate wind patterns that flow toward urban centers at ground level, rise over them, and then move outward aloft.

The UBL can be divided into four vertical layers: the urban canopy layer (UCL), which extends to the average building height; the wake layer, where the individual buildings still significantly influence the airflow; the surface layer; and the Ekman layer. Winds in the UBL differ from those over flat terrain, necessitating extensive field and numerical studies. A better understanding of UBL turbulence is essential for designing civil structures and deploying urban wind turbines, with ongoing efforts to develop comprehensive models that address various spatial and temporal scales.

Wind profiles in the surface layer and Ekman layer over cities can be characterized using the laws discussed in Chap. 4 by selecting a substantial roughness length, typically 1 m or more, along with a displacement height of approximately 2/3 of the average building height. When the power law is applied, a significant power law exponent should also be chosen, usually 0.3 or higher for neutral atmospheric stability conditions.

6.3.1 Atmospheric Stability Over Urban Surface

Figure 6.5 and the subsequent figures show observations of wind and thermodynamic structure over King's Park, which is a high-density urban area in Hong Kong. The observation was performed for a period of one year by a lidar and a microwave radiometer.

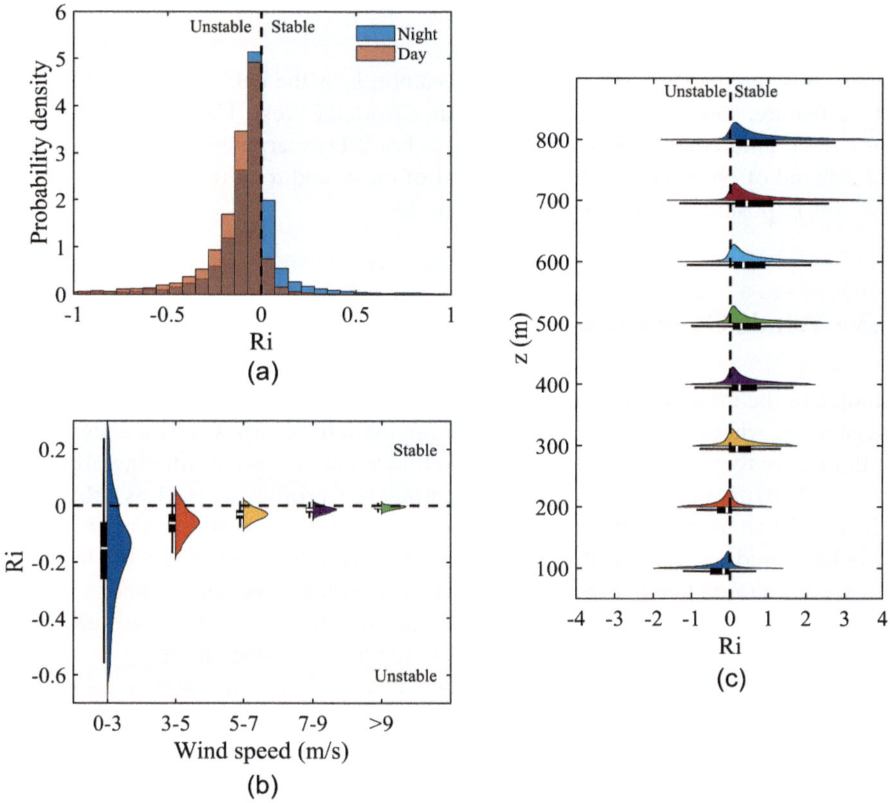

Fig. 6.5 **a** Diurnal variation of surface (0–50 m) bulk Richardson number (Ri), **b** violin plot to show the variation surface Ri with near-surface (50-m-height) wind speed, and **c** violin plot to show the variation of Ri with height over an urban surface (He et al., 2024b)

Figure 6.5a illustrates the diurnal variation in the probability distribution of the bulk Richardson number (Ri) at the surface (0–50 m above ground level). The static stability is indicated by the shaded area. Day represents 0700–1900 LT (local time, UTC + 8), while night represents 1900–2400 and 0000–0700 LT. The results indicate that the ABL at this urban location is predominantly unstable, with an 84.5% probability of negative Ri and a 15.5% probability of positive Ri. This instability is linked to the urban surface characteristics, such as substantial roughness elements, extensive sealed areas, limited moisture availability, and increased heat storage capacity (Rotach et al., 2005). These factors contribute to enhanced sensible heat fluxes from the urban surface into the ABL during the day and diminished radiative cooling at night (Emeis, 2018), leading to a higher likelihood of unstable stratification. In terms of dynamic stability, there is only a 2.1% probability of Ri exceeding 0.25, indicating dynamically stable conditions where flow may intermittently shift between turbulent and laminar states due to thermal suppression of turbulence (Grachev et al., 2013). The likelihood of statically stable conditions at night (23.9%) is approximately three times greater than during the day (7.3%), with nearly all dynamically stable instances occurring at night. This diurnal stability variation can be attributed to thermal mixing from solar heating during the day and thermal suppression from radiative cooling at night. Despite these fluctuations, the stability changes in this urban area are less pronounced than those in rural regions, where the probability of statically stable conditions can exceed 50% at night (Kelly & Gryning, 2010; Peña et al., 2016). This reduced variation is due to the urban surface's high specific heat capacity, which dampens changes in the vertical virtual potential temperature gradient and stability.

Figure 6.5b shows the relationship between the probability distribution of surface Ri and near-surface wind speed (50 m height). As wind speed increases, Ri values cluster around zero, suggesting that near-neutral conditions are prevalent in high wind scenarios. This phenomenon occurs because strong vertical wind shear at elevated speeds generates significant turbulence, overshadowing thermal influences.

Figure 6.5c presents the probability distributions of Ri at various heights. Below 200 m, Ri values are predominantly negative, indicating a high likelihood of statically unstable stratification. The urban surface characteristics contribute to increased daytime sensible heat fluxes and reduced nighttime cooling, leading to a higher probability of negative Ri near the ground. Between 200 and 300 m, a transition occurs from predominantly negative Ri to mostly positive Ri, attributed to the shift from the surface layer to the mixed layer and entrainment zone in the unstable boundary layer. In the surface layer, the lapse rate of virtual potential temperature is negative, becoming approximately zero in the mixed layer and then positive in the entrainment zone. As height increases, the likelihood of positive Ri rises, reflecting a greater occurrence of statically stable layers at higher altitudes, which correlates with stable background stratifications above the ABL. Similarly, the probability of dynamically stable layers with Ri exceeding 0.25 also increases with height, indicating a reduced likelihood of turbulence at greater elevations.

6.3.2 Wind Profiles in the Urban Boundary Layer

Figure 6.6a illustrates wind speed profiles categorized by 50-m height intervals in neutral atmospheric stability conditions. As mentioned before, the UBL is composed of several distinct layers: the roughness sublayer (RSL), which includes the urban canopy layer and wake layer; the inertial sublayer (ISL) or surface layer; and the Ekman layer (EL). Above the Ekman layer, which marks the top of the atmospheric boundary layer (ABL), lies the free atmosphere (FA). Wind speed profiles within the ISL, ranging from 50 to 200 m, generally conform to the logarithmic law, while profiles below 50 m and above 200 m exhibit deviations from this pattern. In the RSL, wind speeds are affected by individual roughness elements, such as buildings and vegetation. Meanwhile, in the EL, Coriolis forces significantly influence both wind speed and direction.

Figure 6.6b depicts wind speed profiles within the lowest 1000 m, with the ABL height for each speed bin marked in black. The profiles reveal the presence of "super-geostrophic" winds or low-level jets just beneath the ABL height. This phenomenon is likely due to the ABL not being "truly neutral" (characterized by zero surface virtual potential temperature flux and zero N^2 in the free atmosphere), but rather "conventionally neutral" (with zero surface flux but positive N^2 in the free atmosphere). Observations indicate that the neutral ABL is typically conventionally neutral, with a free-atmosphere N^2 of approximately $1–3 \times 10^{-4}\ s^{-2}$ and a Zilitinkevich number ranging from 200 to 300. The influence of stably stratified flow in the free atmosphere significantly alters the wind and thermodynamic structure of the conventionally neutral ABL, distinguishing it from a truly neutral ABL (Zilitinkevich & Esau, 2002, 2005; Zilitinkevich et al., 2007). The observed profiles of low-level jet wind speeds closely align with LES results for conventionally neutral ABL, where the jet

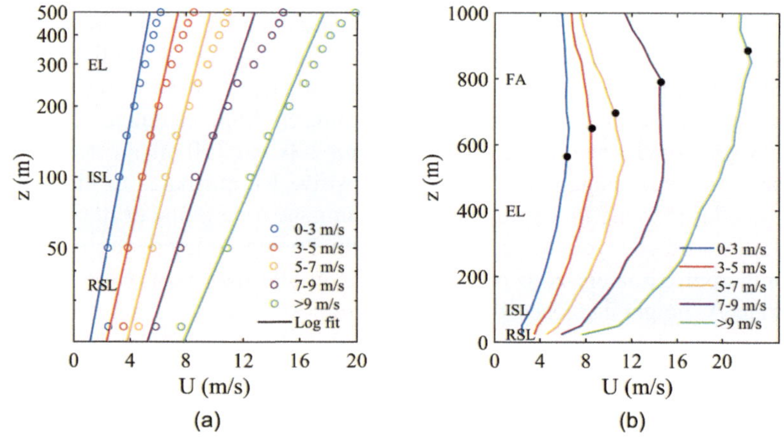

Fig. 6.6 Urban wind speed profiles grouped by 50-m-height wind speed in neutral conditions, **a** over the lowest 500 m in logarithmic height scale, and **b** over the lowest 1000 m (He et al., 2024b)

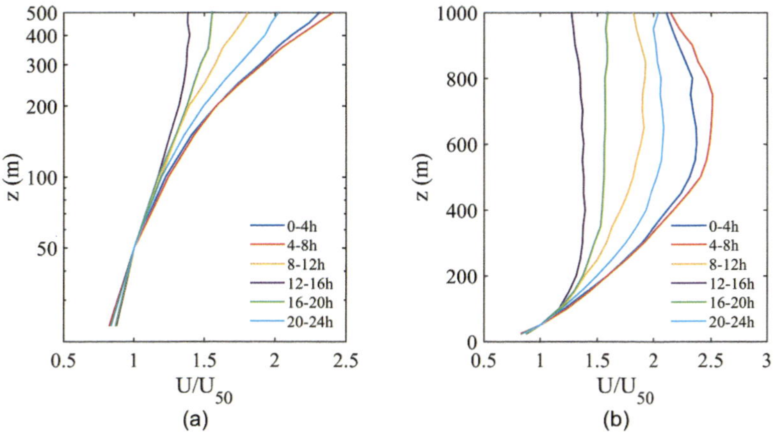

Fig. 6.7 Diurnal variation of urban wind speed profiles **a** over the lowest 500 m in logarithmic height scale, and **b** over the lowest 1000 m (He et al., 2024b)

height is approximately 80–90% of the ABL height (Kelly et al., 2019; Liu et al., 2021).

Figure 6.7 illustrates the diurnal variation of wind speed profiles, while Fig. 6.8a presents the diurnal changes in the nondimensional wind speed gradient $(\partial U/\partial z)/(U/z)$. The observed cycles in wind speed and wind shear correspond closely to the diurnal changes in atmospheric stability depicted in Fig. 6.5a. Below 100 m, the variation in the nondimensional wind speed gradient is less pronounced, as wind shear near the ground is primarily "mechanically driven", influenced by surface roughness. In the 100–500-m range, wind shear becomes more "thermally driven", with clear distinctions between low wind shear during daytime convection and high wind shear at night. The transition to a convective ABL is particularly notable from 7 to 12 h. Above 500 m, wind shear tends to be negligible or even negative at night, reflecting the effects of nocturnal low-level jets.

Additionally, Fig. 6.8b shows the diurnal variation of wind direction veer angles. These angles are generally larger at night compared to the daytime, aligning with the diurnal patterns of atmospheric stability shown in Fig. 6.5a. Wind backing is most pronounced in the afternoon (12–16 h) when conditions are calm and the ABL is convective, while wind veering is most significant late at night (4–8 h) in stable conditions. Previous research indicates that wind veer is influenced not only by ABL stability but also by baroclinicity (Peña et al., 2014).

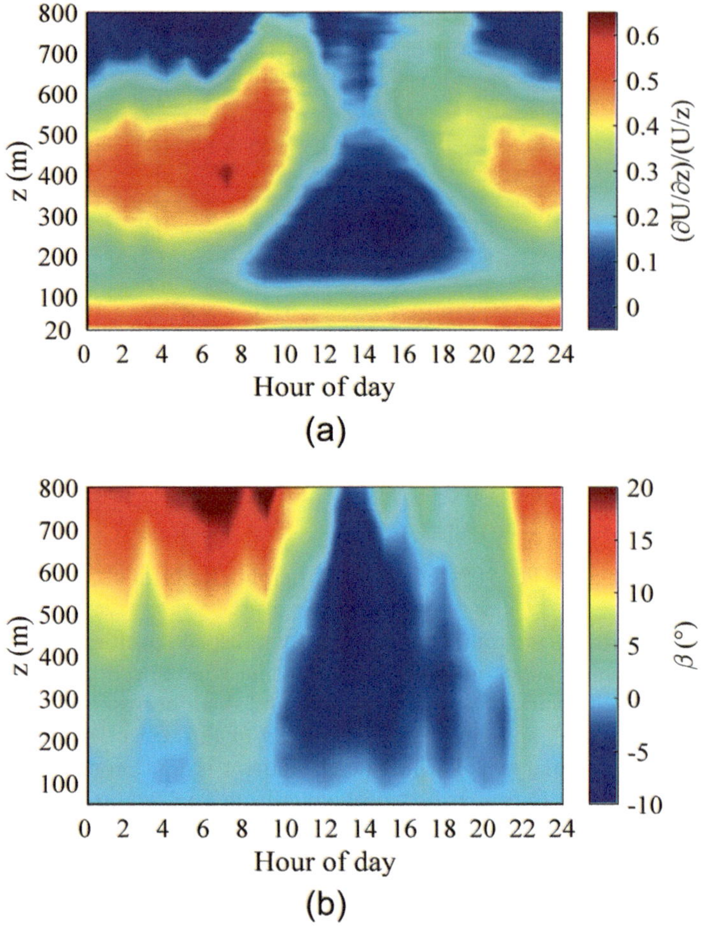

Fig. 6.8 Diurnal variation of **a** nondimensional wind speed gradient (∂U/∂z)/(U/z) and **b** wind direction veer angle at different heights over an urban surface (He et al., 2024b)

6.4 Winds Over Sea

The characteristics of winds over sea are crucial, especially as a growing portion of wind energy is generated at offshore wind farms. While the sea surface is uniformly flat, wind profiles are different from those over homogenous land, as discussed in Chap. 4.

One key difference is that the sea surface is significantly smoother than land, resulting in a higher wind speed and lower turbulence intensity at a given height, as well as a shallower surface layer depth. Moreover, sea surface roughness is influenced by wind speed due to wave formation. Unlike land, diurnal temperature cycles and atmospheric stability variations are minimal over water due to its high heat capacity, which tends to create unstable conditions in the marine ABL.

The marine ABL can be divided into several sublayers. Adjacent to the sea surface is a wave sublayer, where the effects of individual waves dominate, extending to a height of about five wave amplitudes. Above this is the surface layer, which is generally shallower than its land counterpart, often around 10 m in stable conditions and light winds. The upper portion of the marine ABL is characterized by the Ekman layer. Overall, the marine ABL is typically much shallower than the ABL over land.

6.4.1 Sea Surface Roughness and Wind Profiles

The roughness length of the sea surface at moderate wind speeds ranges from 0.1 to 1 mm, with a power law exponent of about 0.1 for neutral atmospheric stability conditions. The sea surface roughness varies significantly with wind speed due to changes in wave size and shape, increasing with stronger winds. Waves form primarily from frictional forces between the wind and the ocean surface, transferring momentum downward into the water column. This downward transport occurs as long as the wind speed exceeds the phase speed of the waves. For older waves or swells, the relationship with wind speed becomes less clear. In addition, the thermal state of the marine ABL affects this momentum transport. In unstable conditions, downward transport is greater, leading to taller waves compared to stable conditions.

One of the earliest parameterizations of z_0 of the sea surface is the Charnock's relation (Charnock, 1955):

$$z_0 = \alpha_c \frac{u_*^2}{g} \tag{6.8}$$

where $\alpha_c \cong 0.015$ is the Charnock parameter. However, this relation underestimates z_0 for low wind speeds, because under these circumstances existing capillary waves are very rough. It is therefore recommended to use the relation by Zilitinkevich (1969):

$$z_0 = \frac{c_1 \nu}{u_*} + \frac{u_*^2}{c_2 g} \tag{6.9}$$

where ν is the kinematic viscosity of air ($\nu \cong 1.47 \times 10^{-5}$ m²/s at $T = 15$ °C), c_1 and c_2 are coefficients suggested as $c_1 = 0.1$ and $c_2 = 20.8$ (Zilitinkevich, 1969), $c_1 = 0.11$ and $c_2 = 90.9$ (Smith, 1988), $c_1 = 0.11$ and $c_2 = 55.6$ (Beljaars, 1995), and $c_1 = 0.11$ and $c_2 = 56$ (for open ocean) or 32 (for coastal zone) (Zilitinkevich et al., 2002), $c_1 = 0.48$ and $c_2 = 81.1$ (Foken, 2017).

Other studies that measured the 10-m drag coefficient C_d (defined as $C_d = (u_*/U_{10})^2$, where U_{10} is the wind speed at 10-m height) for neutral conditions also provided formulae for calculating z_0, e.g., Large and Pond (1981):

$$z_0 = \frac{10}{\exp\left(k/\sqrt{C_d}\right)} \tag{6.10}$$

$$C_d = \begin{cases} 1.2 \times 10^{-3}, & U_{10} \leq 11\,\text{m/s} \\ (0.49 + 0.065U_{10}) \times 10^{-3}, & U_{10} \geq 11\,\text{m/s} \end{cases} \tag{6.11}$$

Recent field studies for high wind speeds in TCs indicated that the drag coefficient and roughness length level off as the wind speeds increase above hurricane force of 33 m/s (Powell et al., 2003). This is also supported by the water tank experiment of Donelan et al. (2004), which suggested the following variation of C_d with U_{10}:

$$C_d = \begin{cases} 0.6 \times 10^{-3}, & U_{10} \leq 5\,\text{m/s} \\ (0.25 + 0.065U_{10}) \times 10^{-3}, & 5 \leq U_{10} \leq 33\,\text{m/s} \\ 2.4 \times 10^{-3}, & U_{10} \geq 33\,\text{m/s} \end{cases} \tag{6.12}$$

Variations in drag coefficients and roughness lengths reported in different studies are likely influenced by the state of the sea, including factors like wave steepness and age. Research indicates that roughness lengths tend to increase with younger waves, meaning they decrease with wave age. Understanding the specific relationship between the roughness lengths and these factors remains an active area of investigation in air-sea interaction studies.

The marine wind structure is closely linked to wind speed, which affects wave behavior and sea state through air-sea interactions, thereby influencing the wind profile above the ocean surface. Figure 6.9 illustrates the relationships between the wind shear coefficient, wind veer rate, roughness length, and friction velocity with 100-m wind speed for a neutral marine ABL over shallow coastal waters. The observation was performed by a lidar on a marine platform in the Southeastern Waters of Hong Kong.

Figure 6.9a shows the variation of wind shear coefficient with wind speed. The dots represent median values, while the error bars indicate the 25th and 75th percentiles. α tends to increase with wind speed, which correlates with the increase in z_0 depicted in Fig. 6.9c. Notably, α at lower heights is typically higher than that at greater heights, consistent with theoretical expectations.

In Fig. 6.9b, the wind veer rate β is observed to decrease as the 100-m wind speed increases. This decline cannot be explained by the geostrophic drag law, which suggests that higher wind speed should elevate both u_* and z_0, reduce the friction Rossby number, and increase the cross-isobar angle. Instead, the decrease in β is likely due to enhanced mechanical mixing and greater heights of the surface and boundary layers at elevated wind speeds (Tsai et al., 2019; Zhao et al., 2022). Other potential factors include background stratification (Kelly et al., 2019) and baroclinicity (Ghannam & Bou-Zeid, 2021).

Figure 6.9c, d show how z_0 and u_* vary with hub-height wind speed. These values were determined through least-squares fitting of wind profiles over two height ranges

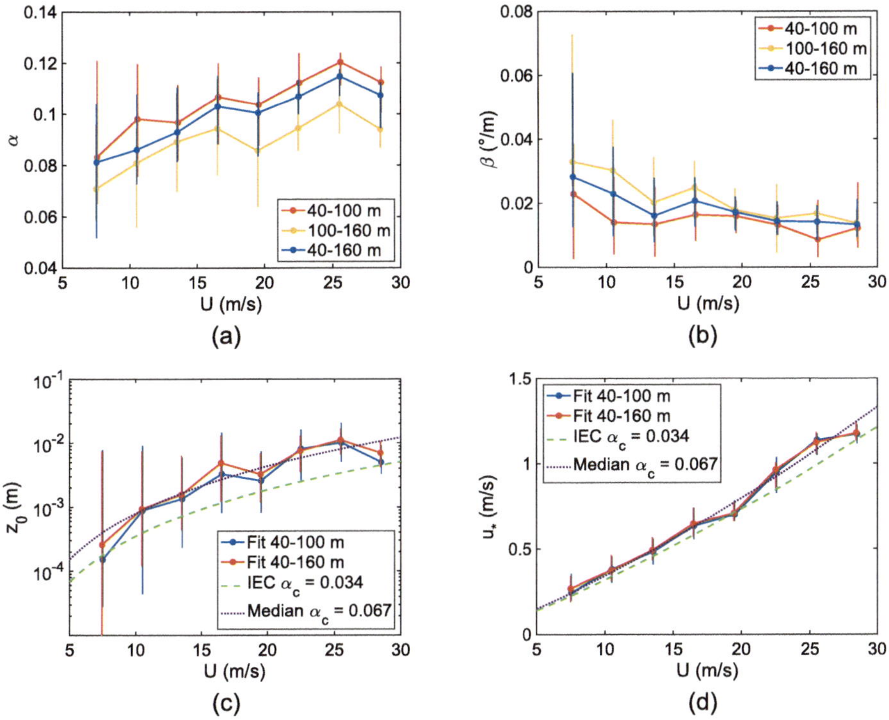

Fig. 6.9 Variation of **a** wind shear coefficient, **b** wind veer rate, **c** roughness length, and **d** friction velocity with wind speed for a marine atmospheric boundary layer (He et al., 2024a)

to the neutral logarithmic law, revealing that both z_0 and u_* generally increase with wind speed. However, the observed z_0 exceeds that predicted by the Charnock relationship, which suggests a coefficient $\alpha_c = 0.034$ for near-coastal areas, as recommended by IEC (2019). To find an "optimal" Charnock coefficient for this site, α_c is calculated as $\alpha_c = z_0 g/u_*^2$ (Edson et al., 2013), resulting in a median α_c of 0.067. The elevated α_c and z_0 in this study are likely related to site-specific characteristics, such as shallow water depth, which may cause increased wave steepness (Taylor & Yelland, 2001; Zhao et al., 2015). Previous studies, such as Johnson et al. (1988), have also reported high α_c values exceeding 0.1 in shallow coastal waters.

Furthermore, it is observed that both z_0 and α tend to decrease with increasing wind speed when 100-m wind speed > 25 m/s. This trend may be linked to wave breaking and sea spray at high wind speeds (Takagaki et al., 2016). Wave breaking is likely to occur at lower wind speeds in shallow coastal waters (around 25 m/s) compared to deeper oceans (30–40 m/s) (Bi et al., 2015; Holthuijsen et al., 2012; Powell et al., 2003; Zhao et al., 2015), which aligns with the current findings.

Figure 6.10 shows the wind profiles for different bins of 100-m wind speed. For each wind speed bin, the ensemble wind profile is constructed based on the median of 10-min mean wind speed profile samples. In general, there is no apparent discrepancy

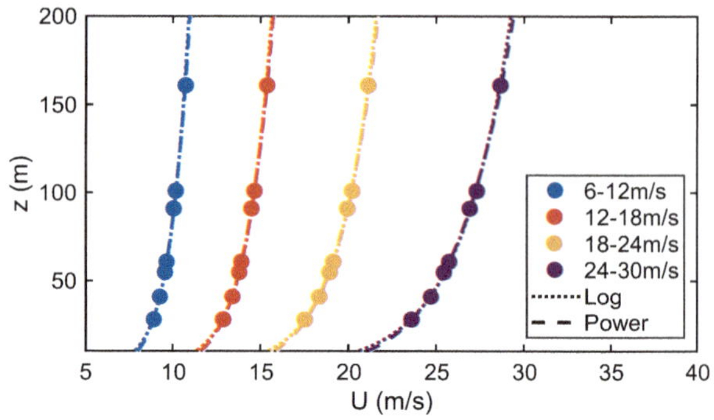

Fig. 6.10 Variation of wind profiles with wind speed for a marine atmospheric boundary layer (He et al., 2024a)

between the profiles fitted by the logarithmic law and the power law, and both laws describe the ensemble wind profiles excellently ($R^2 > 0.99$).

6.4.2 Turbulence and Gusts Over Sea

Figure 6.11 illustrates the variation of the longitudinal turbulence intensity and its vertical profile in relation to 100-m wind speed. Overall, turbulence intensity increases with wind speed but begins to plateau or decline when wind speed exceeds 25 m/s, consistent with the trends observed in roughness length shown in Fig. 6.9c. The power law effectively describes the turbulence intensity profiles, achieving a high correlation coefficient of 0.97–0.99.

Figure 6.12 illustrates the relationship between longitudinal gust factors and mean wind speeds over a sea surface. It shows that for 1-min mean wind speeds $U(60\text{ s})$ ranging from 10 to 70 m/s, the gust factor $G(3\text{ s}, 60\text{ s})$ increases from 1.14 at $U(60\text{ s}) = 10$ m/s to 1.2 at $U(60\text{ s}) = 30$ m/s, then levels off between 30 and 50 m/s before decreasing to 1.13 at $U(60\text{ s}) = 70$ m/s. The variability in observed gust factors may

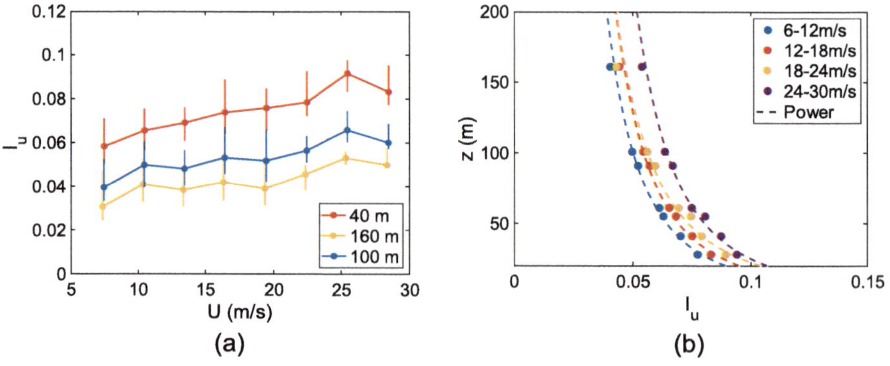

Fig. 6.11 Variation of **a** longitudinal turbulence intensity and **b** longitudinal turbulence intensity profiles with wind speed for a marine atmospheric boundary layer (He et al., 2024a)

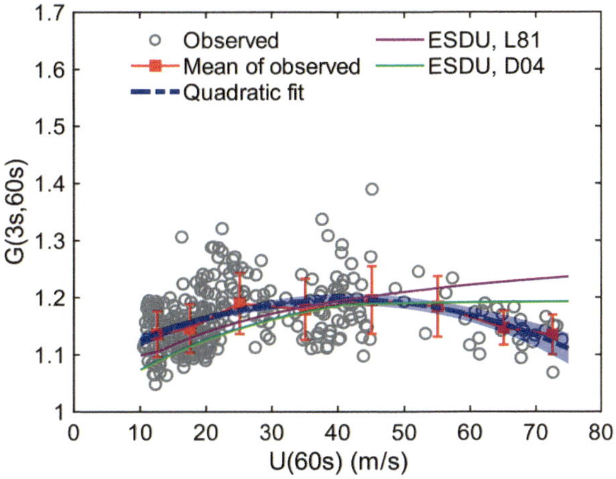

Fig. 6.12 Variation of gust factor with wind speed over the sea (He et al., 2021)

be linked to the transient and turbulent nature of tropical cyclone winds, as well as thermally generated turbulence influenced by atmospheric stability (Shu et al., 2015).

The decline in gust factor at very high wind speeds may be attributed to reduced sea surface roughness due to breaking waves and sea spray. Roughness length z_0 is correlated with mean wind speed U, as wave characteristics depend on this speed (Reichl et al., 2014). Earlier research indicated a consistent increase in z_0 with U (Large & Pond, 1981; Smith, 1988). However, recent studies have shown that z_0 levels off or even decreases at wind speeds above 30 m/s due to the effects of breaking waves and sea spray (Edson et al., 2013; Jarosz et al., 2007; Kudryavtsev & Makin, 2007; Powell et al., 2003; Troitskaya et al., 2012). Under near-neutral stability during strong tropical cyclone winds, terrain exposure becomes the main factor affecting the near-surface gust factor, which generally increases with z_0 (Giammanco et al., 2016; Miller et al., 2015). Thus, as mean wind speed increases, sea surface roughness decreases, leading to lower gust factors. Holthuijsen et al. (2012) suggested that z_0 approaches zero as mean wind speeds near 80 m/s, resulting in an aerodynamically smooth surface, which aligns with the observed decrease in gust factors in Fig. 6.12. It is hypothesized that at extreme wind speeds exceeding 80 m/s, the gust factor may approach one, indicating minimal mechanical turbulence due to a very smooth sea surface, although this requires further investigation.

Another potential explanation for the observed phenomena could be turbulence generated by convective processes in the eyewalls and outer vortex of tropical cyclones. However, Giammanco et al. (2016) found minimal differences in mean gust factors between the eyewall and outer vortex, suggesting that convective turbulence is not responsible for the decrease in gust factors at high wind speeds.

To assess whether the observed gust factors reflect the ocean surface conditions, the expected mean gust factors at heights of 66.5 m above mean sea level were calculated and are shown in Fig. 6.12. The gust factors at 66.5 m, predicted by the ESDU model using the Large and Pond (1981) and Donelan et al. (2004) roughness parameterizations, are represented by the magenta and green solid lines, respectively.

The observed G(3 s, 60 s) values align closely with those predicted by the ESDU model at 66.5 m. For lower wind speeds (below 30 m/s), the behavior of G(3 s, 60 s) closely resembles predictions from both roughness parameterizations, although observed values are slightly higher, likely due to topographic effects. At wind speeds between 30 and 50 m/s, the plateau in observed G(3 s, 60 s) corresponds with the predictions from the Donelan et al. (2004) parameterization, indicating a capped gust factor in this regime. In contrast, the continuously increasing gust factors suggested by the ESDU model using the Large and Pond (1981) parameterization diverge from the observed data. A notable finding from this study is the decrease in G(3 s, 60 s) at wind speeds exceeding 50 m/s, although limited observations exist for such high speeds. Cao et al. (2015) reported a significant reduction in gust factors during a strong typhoon when mean wind speeds surpassed 30 m/s, and Fang et al. (2019) observed a slight decrease at 10-min mean wind speeds above 32.5 m/s from measurements taken at coastal meteorological towers.

References

Beljaars, A. C. M. (1995). The parametrization of surface fluxes in large-scale models under free convection. *Quarterly Journal of the Royal Meteorological Society, 121*(522), 255–270. https://doi.org/10.1002/qj.49712152203

Bi, X., Gao, Z., Liu, Y., Liu, F., Song, Q., Huang, J., Huang, H., Mao, W., & Liu, C. (2015). Observed drag coefficients in high winds in the near offshore of the South China Sea. *Journal of Geophysical Research Atmospheres, 120*, 6444–6459. https://doi.org/10.1002/2015JD023172

Cao, S., Tamura, Y., Kikuchi, N., Saito, M., Nakayama, I., & Matsuzaki, Y. (2015). A case study of gust factor of a strong typhoon. *Journal of Wind Engineering & Industrial Aerodynamics, 138*, 52–60. https://doi.org/10.1016/j.jweia.2014.12.012

Charnock, H. (1955). Wind stress on a water surface. *Quarterly Journal of the Royal Meteorological Society, 81*, 639–640. https://doi.org/10.1002/qj.49708135027

Donelan, M. A., Haus, B. K., Reul, N., Plant, W. J., Stiassnie, M., Graber, H. C., Brown, O. B., & Saltzman, E. S. (2004). On the limiting aerodynamic roughness of the ocean in very strong winds. *Geophysical Research Letters, 31*, L18306. https://doi.org/10.1029/2004GL019460

Edson, J. B., Jampana, V., Weller, R. A., Bigorre, S. P., Plueddemann, A. J., Fairall, C. W., Miller, S. D., Mahrt, L., Vickers, D., & Hersbach, H. (2013). On the exchange of momentum over the open ocean. *Journal of Physical Oceanography, 43*, 1589–1610. https://doi.org/10.1175/JPO-D-12-0173.1

Elliott, W. P. (1958). The growth of the atmospheric internal boundary layer. *Eos, Transactions American Geophysical Union, 39*, 1048–1054. https://doi.org/10.1029/TR039i006p01048

Emeis, S. (2018). *Wind energy meteorology—Atmospheric physics for wind power generation* (2nd ed.). Springer.

Fang, G., Zhao, L., Cao, S., Ge, Y., & Li, K. (2019). Gust characteristics of near-ground typhoon winds. *Journal of Wind Engineering & Industrial Aerodynamics, 188*, 323–337. https://doi.org/10.1016/j.jweia.2019.03.008

Foken, T. (2017). *Micrometeorology* (2nd ed.). Springer.

Ghannam, K., & Bou-Zeid, E. (2021). Baroclinicity and directional shear explain departures from the logarithmic wind profile. *Quarterly Journal of the Royal Meteorological Society, 147*(734), 443–464. https://doi.org/10.1002/qj.3927

Giammanco, I. M., Schroeder, J. L., Masters, F. J., Vickery, P. J., Krupar, R. J., & Balderrama, J. A. (2016). Influences on observed near-surface gust factors in landfalling U.S. Gulf coast

hurricanes: 2004–08. *Journal of Applied Meteorology and Climatology, 55*, 2587–2611. https://doi.org/10.1175/JAMC-D-16-0053.1
Grachev, A. A., Andreas, E. L., Fairall, C. W., Guest, P. S., & Persson, P. O. G. (2013). The critical Richardson number and limits of applicability of local similarity theory in the stable boundary layer. *Boundary-Layer Meteorology, 147*, 51–82. https://doi.org/10.1007/s10546-012-9771-0
He, J. Y., Chan, P. W., & Li, Q. S. (2024a). Statistical characterization of marine wind structure over wind-turbine-relevant heights in tropical cyclones. *Journal of Wind Engineering and Industrial Aerodynamics, 253*, 105874. https://doi.org/10.1016/j.jweia.2024.105874
He, J. Y., Chan, P. W., Liu, Z. F., Fung, J. C. H., & Li, Q. S. (2024b). Observation of wind and thermodynamic structure within an urban boundary layer. *Physics of Fluids, 36*(8), 086603. https://doi.org/10.1063/5.0214961
He, J. Y., Li, Q. S., & Chan, P. W. (2021). Reduced gust factor for extreme tropical cyclone winds over ocean. *Journal of Wind Engineering & Industrial Aerodynamics, 208*, 104445. https://doi.org/10.1016/j.jweia.2020.104445
Holmes, J. D. (2015). *Wind loading of structures* (3rd ed.). CRC Press.
Holthuijsen, L. H., Powell, M. D., & Pietrzak, J. D. (2012). Wind and waves in extreme hurricanes. *Journal of Geophysical Research: Oceans, 117*, 1–15. https://doi.org/10.1029/2012JC007983
IEC (2019). IEC61400-3-1:2019 Wind Energy Generation Systems - Part 3-1: Design Requirements for Fixed Offshore Wind Turbines.
Johnson, H. K., Højstrup, J., Vested, H. J., & Larsen, S. E. (1998). On the dependence of sea surface roughness on wind waves. *Journal of Physical Oceanography, 28*(9), 1702–1716. https://doi.org/10.1175/1520-0485(1998)0282.0.CO;2
Jarosz, E., Mitchell, D. A., Wang, D. W., & Teague, W. J. (2007). Bottom-up determination of air-sea momentum exchange under a major tropical cyclone. *Science, 315*, 1707–1709. https://doi.org/10.1126/science.1136466
Kelly, M., Cersosimo, R. A., & Berg, J. (2019). A universal wind profile for the inversion-capped neutral atmospheric boundary layer. *Quarterly Journal of the Royal Meteorological Society, 145*, 982–992. https://doi.org/10.1002/qj.3472
Kelly, M., & Gryning, S. E. (2010). Long-term mean wind profiles based on similarity theory. *Boundary-Layer Meteorology, 136*, 377–390. https://doi.org/10.1007/s10546-010-9509-9
Kudryavtsev, V. N., & Makin, V. K. (2007). Aerodynamic roughness of the sea surface at high winds. *Boundary-Layer Meteorology, 125*, 289–303. https://doi.org/10.1007/s10546-007-9184-7
Large, W. G., & Pond, S. (1981). Open ocean momentum flux measurements in moderate to strong winds. *Journal of Physical Oceanography, 11*, 324–336. https://doi.org/10.1175/1520-0485(1981)011%3c0324:oomfmi%3e2.0.co;2
Liu, L., Gadde, S. N., & Stevens, R. J. A. M. (2021). Universal wind profile for conventionally neutral atmospheric boundary layers. *Physical Review Letters, 126*, 104502. https://doi.org/10.1103/PhysRevLett.126.104502
Logan, E., & Fichtl, G. H. (1975). Rough-to-smooth transition of an equilibrium neutral constant stress layer. *Boundary-Layer Meteorology, 8*(3), 525–528. https://doi.org/10.1007/BF02153569
Miller, C., Balderrama, J. A., & Masters, F. (2015). Aspects of observed gust factors in landfalling tropical cyclones: Gust components, terrain, and upstream fetch effects. *Boundary-Layer Meteorology, 155*, 129–155. https://doi.org/10.1007/s10546-014-9989-0
Peña, A., Floors, R., & Gryning, S. E. (2014). The Høvsøre tall wind-profile experiment: A description of wind profile observations in the atmospheric boundary layer. *Boundary-Layer Meteorology, 150*, 69–89. https://doi.org/10.1007/s10546-013-9856-4
Peña, A., Floors, R., Sathe, A., Gryning, S. E., Wagner, R., Courtney, M. S., Larsén, X. G., Hahmann, A. N., & Hasager, C. B. (2016). Ten years of boundary-layer and wind-power meteorology at Høvsøre, Denmark. *Boundary-Layer Meteorology, 158*, 1–26. https://doi.org/10.1007/s10546-015-0079-8
Powell, M. D., Vickery, P. J., & Reinhold, T. A. (2003). Reduced drag coefficient for high wind speeds in tropical cyclones. *Nature, 402*, 279–283. https://doi.org/10.1038/nature01481
Reichl, B. G., Hara, T., & Ginis, I. (2014). Sea state dependence of the wind stress over the ocean under hurricane winds. *Journal of Geophysical Research: Oceans, 119*, 30–51. https://doi.org/10.1002/2013JC009289

Rotach, M. W., Vogt, R., Bernhofer, C., Batchvarova, E., Christen, A., Clappier, A., Feddersen, B., Gryning, S. E., Martucci, G., Mayer, H., Mitev, V., Oke, T. R., Parlow, E., Richner, H., Roth, M., Roulet, Y. A., Ruffieux, D., Salmond, J. A., Schatzmann, M., & Voogt, J. A. (2005). BUBBLE—An urban boundary layer meteorology project. *Theoretical and Applied Climatology, 81*(3–4), 231–261. https://doi.org/10.1007/s00704-004-0117-9

Shu, Z. R., Li, Q. S., He, Y. C., & Chan, P. W. (2015). Gust factors for tropical cyclone, monsoon and thunderstorm winds. *Journal of Wind Engineering & Industrial Aerodynamics, 142*, 1–14. https://doi.org/10.1016/j.jweia.2015.02.003

Smith, S. D. (1988). Coefficients for sea surface wind stress, heat flux, and wind profiles as a function of wind speed and temperature. *Journal of Geophysical Research: Oceans, 93*, 15467–15472. https://doi.org/10.1029/JC093iC12p15467

Takagaki, N., Komori, S., Suzuki, N., Iwano, K., & Kurose, R. (2016). Mechanism of drag coefficient saturation at strong wind speeds. *Geophysical Research Letters, 43*, 9829–9835. https://doi.org/10.1002/2016GL070666

Taylor, P. K., & Yelland, M. J. (2001). The dependence of sea surface roughness on the height and steepness of the waves. *Journal of Physical Oceanography, 31*, 572–590. https://doi.org/10.1175/1520-0485(2001)0312.0.CO;2

Troitskaya, Y. I., Sergeev, D. A., Kandaurov, A. A., Baidakov, G. A., Vdovin, M. A., & Kazakov, V. I. (2012). Laboratory and theoretical modeling of air-sea momentum transfer under severe wind conditions. *Journal of Geophysical Research: Oceans, 117*, 1–13. https://doi.org/10.1029/2011JC007778

Tsai, Y.-S., Miau, J.-J., Yu, C.-M., & Chang, W.-T. (2019). Lidar observations of the typhoon boundary layer within the outer rainbands. *Boundary-Layer Meteorology, 171*, 237–255. https://doi.org/10.1007/s10546-019-00427-6

Wood, D. H. (1982). Internal boundary layer growth following a step change in surface roughness. *Boundary-Layer Meteorology, 22*, 241–244. https://doi.org/10.1007/BF00118257

Yan, B. W., & Li, Q. S. (2016). Coupled on-site measurement/CFD based approach for high-resolution wind resource assessment over complex terrains. *Energy Conversion and Management, 117*, 351–356. https://doi.org/10.1016/j.enconman.2016.02.076

Zhao, Z., Gao, R., Zhang, J. A., Zhu, Y., Liu, C., Chan, P. W., & Wan, Q. (2022). Observations of boundary layer wind and turbulence of a landfalling tropical cyclone. *Scientific Reports, 12*(1), 11056. https://doi.org/10.1038/s41598-022-14929-w

Zhao, Z.-K., Liu, C.-X., Li, Q., Dai, G.-F., Song, Q.-T., & Lv, W.-H. (2015). Typhoon air-sea drag coefficient in coastal regions. *Journal of Geophysical Research: Oceans, 120*, 716–727. https://doi.org/10.1002/jgrc.20224

Zilitinkevich, S. S. (1969). On the computation of the basic parameters of the interaction between the atmosphere and the ocean. *Tellus, 21*.

Zilitinkevich, S. S., Esau, I., & Baklanov, A. (2007). Further comments on the equilibrium height of neutral and stable planetary boundary layers. *Quarterly Journal of the Royal Meteorological Society, 133*, 265–271. https://doi.org/10.1002/qj.27

Zilitinkevich, S. S., & Esau, I. N. (2002). On integral measures of the neutral barotropic planetary boundary layer. *Boundary-Layer Meteorology, 104*, 371–379. https://doi.org/10.1023/A:1016540808958

Zilitinkevich, S. S., & Esau, I. N. (2005). Resistance and heat-transfer laws for stable and neutral planetary boundary layers: Old theory advanced and re-evaluated. *Quarterly Journal of the Royal Meteorological Society, 131*, 1863–1892. https://doi.org/10.1256/qj.04.143

Zilitinkevich, S. S., Perov, V. L., & King, J. C. (2002). Near-surface turbulent fluxes in stable stratification: Calculation techniques for use in general-circulation models. *Quarterly Journal of the Royal Meteorological Society, 128*(583), 1571–1587. https://doi.org/10.1002/qj.200212858309

Chapter 7
Wind Climatology

The previous chapters focus on the micrometeorology of winds in the ABL, while this chapter provides a brief introduction to the wind climatology. Topics mainly include probability distribution of wind speed (Sect. 7.1), long-term mean wind profile (Sect. 7.2), macroscale and mesoscale wind spectrum (Sect. 7.3), and climate change (Sect. 7.4).

7.1 Probability Distribution of Wind Speed

In view of the cubic relationship between the wind speed and wind energy, comprehensive and accurate information of the wind speed is required for the best practices of wind resource assessment. However, the wind is highly variable, both geographically and temporally, over a very wide range of scales. While the detailed variation in wind speeds over a long period (e.g., over a year) remains hard to predict, wind speeds and resources can be well characterized in terms of probability distributions.

7.1.1 Weibull Distribution

Conventionally, the two-parameter Weibull distribution function is used to empirically describe the statistical features of wind speed (Chang, 2011a; Gil Ruiz et al., 2021; Li et al., 2016; Mary, 2021; Samal, 2021; Wais, 2017; Yang et al., 2020). The collection of typically averaged (e.g., 10-min, 30-min, or 1-h mean) wind speed observations over a long period at a given height is assumed to follow the two-parameter Weibull distribution, as follows (Chang, 2011a):

$$f_{wei}(x; k, c) = \frac{k}{c}\left(\frac{x}{c}\right)^{k-1} \exp\left[-\left(\frac{x}{c}\right)^{k}\right] \tag{7.1}$$

J. He and Q.-S. Li, *Winds in the Atmospheric Boundary Layer*,
https://doi.org/10.1007/978-981-95-1517-2_7

$$F_{wei}(x;k,c) = 1 - \exp\left[-\left(\frac{x}{c}\right)^k\right] \tag{7.2}$$

where f_{wei} and F_{wei} are the probability density function (PDF) and cumulative distribution function (CDF) of the Weibull distribution, respectively, and k and c are Weibull shape and scale parameters, respectively. k reflects the skewness of the distribution, and c describes the location of the distribution. Figure 7.1 shows the Weibull distribution with different shape and scale parameters.

A variety of numerical methods have been proposed for the estimation of k and c, such as the empirical method (Justus et al., 1978):

$$k = \left(\frac{\sigma}{\mu}\right)^{-1.086} \tag{7.3}$$

$$c = \frac{\mu}{\Gamma(1 + 1/k)} \tag{7.4}$$

where $\mu = \overline{x}$ and $\sigma = \sqrt{\overline{x^2}}$ are the mean and standard deviation of the samples, respectively, and Γ is the gamma function given by:

$$\Gamma(x) = \int_0^\infty t^{x-1} \exp(-t)dt \tag{7.5}$$

Another simple and easy-to-use method is the graphical method (Chang, 2011b). Taking a double logarithmic transformation to the Weibull cumulative distribution function yields:

$$\ln\{-\ln[1 - F(x)]\} = k\ln(x) - k\ln(c) \tag{7.6}$$

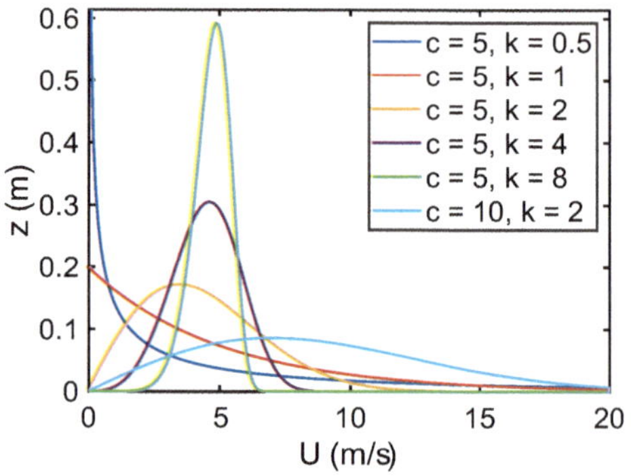

Fig. 7.1 Weibull distribution with different shape and scale parameters

Plot $\ln(x)$ against $\ln\{-\ln[1-F(x)]\}$ and fit a straight line, values of k and c can be estimated based on the slope and intercept of the line.

The moment method is based on the numerical iteration of the following two equations (Chang, 2011b):

$$\mu = c\Gamma(1 + 1/k) \tag{7.7}$$

$$\sigma = c\sqrt{\Gamma(1 + 2/k) - \Gamma^2(1 + 1/k)} \tag{7.8}$$

The maximum likelihood estimation (MLE) method is based on numerical iterations of the following equations (Alexandre et al., 2012; Mohammadi et al., 2016):

$$k = \left[\frac{\sum_{i=1}^{n} x_i^k \ln(x_i)}{\sum_{i=1}^{n} x_i^k} - \frac{\sum_{i=1}^{n} \ln(x_i)}{n}\right]^{-1} \tag{7.9}$$

$$c = \left(\frac{1}{n}\sum_{i=1}^{n} x_i^k\right)^{\frac{1}{k}} \tag{7.10}$$

where x_i, $i = 1, 2, \ldots, n$ are the data samples.

If data is in frequency distribution format, the modified maximum likelihood method can be considered (Chang, 2011b):

$$k = \left[\frac{\sum_{i=1}^{n} x_i^k \ln(x_i) f(x_i)}{\sum_{i=1}^{n} x_i^k f(x_i)} - \sum_{i=1}^{n} \ln(x_i) f(x_i)\right]^{-1} \tag{7.11}$$

$$c = \left(\sum_{i=1}^{n} x_i^k f(x_i)\right)^{\frac{1}{k}} \tag{7.12}$$

Akdağ and Dinler (2009) proposed the following power density method for estimating Weibull parameters for wind energy applications:

$$k = 1 + \frac{3.69}{\left[\overline{x^3}/\mu^3\right]^2} \tag{7.13}$$

$$c = \frac{\mu}{\Gamma(1 + 1/k)} \tag{7.14}$$

To examine the suitability of the probability distribution models in fitting the wind speed observations, the Kolmogorov–Smirnov statistic, which is defined as the maximum distance between the cumulative distribution functions of the sample and the reference probability distribution, can be computed (Amirinia et al., 2017;

Smirnov, 1948):

$$KS = \max|g(U) - f(U)| \tag{7.15}$$

where *KS* is the Kolmogorov–Smirnov statistic, $g(U)$ is the empirical distribution function of the sample (wind speed observations), and $f(U)$ is the cumulative distribution function of the reference distribution. Generally, the smaller the Kolmogorov–Smirnov statistic, the better the reference distribution represents the observation.

7.1.2 Other Unimodal Distributions

Apart from the two-parameter Weibull distribution, other distributions have also been applied in modeling wind speed distribution. One example is the two-parameter Gaussian distribution (also known as the normal distribution):

$$f_{nor}(x; \mu, \sigma) = \frac{1}{\sqrt{2\pi}\sigma} \exp\left[-\frac{(x-\mu)^2}{2\sigma^2}\right] \tag{7.16}$$

$$F_{nor}(x; \mu, \sigma) = \frac{1}{2}\left[1 + \mathrm{erf}\left(\frac{x-\mu}{\sqrt{2}\sigma}\right)\right] \tag{7.17}$$

where μ and σ are the mean and standard deviation of the sample, respectively, and erf is given as:

$$\mathrm{erf}(x) = \frac{2}{\sqrt{\pi}} \int_0^x \exp(-t^2)dt \tag{7.18}$$

The two-parameter lognormal distribution can also be used to model the wind speed distribution (Alavi et al., 2016; Wu et al., 2013):

$$f_{logn}(x; \mu, \sigma) = \frac{1}{x\sigma\sqrt{2\pi}} \exp\left[-\frac{(\ln x - \mu)^2}{2\sigma^2}\right] \tag{7.19}$$

$$F_{logn}(x; \mu, \sigma) = \frac{1}{2}\left[1 + \mathrm{erf}\left(\frac{\ln x - \mu}{\sqrt{2}\sigma}\right)\right] \tag{7.20}$$

Another distribution model is the two-parameter gamma distribution (Carta et al., 2009; Morgan et al., 2011):

$$f_{gam}(x; a, b) = \frac{1}{b^a\Gamma(a)} x^{a-1} \exp\left(-\frac{x}{b}\right) \tag{7.21}$$

$$F_{gam}(x; a, b) = \frac{1}{\Gamma(a)} \gamma\left(a, \frac{x}{b}\right) \tag{7.22}$$

where a and b are shape and scale parameters for the gamma distribution, respectively, γ is the lower incomplete gamma function given by:

$$\gamma(x, s) = \int_0^s t^{x-1} \exp(-t) dt \tag{7.23}$$

In addition to the two-parameter family, there are more flexible distributions with more than two parameters, such as the four-parameter Kappa distribution (Hosking & Wallis, 1997):

$$f_{kap}(x; \xi, \alpha, k, h) = \alpha^{-1} \left[1 - k\left(\frac{x - \xi}{\alpha}\right)\right]^{1/k - 1} \left[F_{kap}(x)\right]^{1-h} \tag{7.24}$$

$$F_{kap}(x; \xi, \alpha, k, h) = \left(1 - h\left[1 - k\left(\frac{x - \xi}{\alpha}\right)\right]^{1/k}\right)^{1/h} \tag{7.25}$$

where ξ, α, k, and h are the location, scale, first shape, and second shape parameters, respectively.

The five-parameter Wakeby distribution is defined by its quantile function (Hosking & Wallis, 1997):

$$F_{wak}^{-1}(p; \xi, \alpha, \beta, \gamma, \delta) = \xi + \frac{\alpha}{\beta}\left[1 - (1 - p)^{\beta}\right] - \frac{\gamma}{\delta}\left[1 - (1 - p)^{-\delta}\right] \tag{7.26}$$

where ξ, α, β, γ, and δ are the location, first scale, first shape, second scale, and second shape parameters, respectively.

Figure 7.2 shows the fit of wind speed observations with Weibull, normal, lognormal, gamma, Kappa, and Wakeby distributions. The wind speed was observed by a lidar at a height of 120 m at an urban site for a period of two years.

7.1.3 Bimodal Distributions

There are wind regimes showing bimodal patterns, which are best described by distribution functions mixed with two or more distributions. One example of bimodal distribution functions is the normal–Weibull mixture distribution (Carta et al., 2009):

$$f_{N-W}(x; \omega, \mu, \sigma, k, c) = \omega \frac{1}{\sqrt{2\pi}\sigma} \exp\left[-\frac{(x - \mu)^2}{2\sigma^2}\right]$$

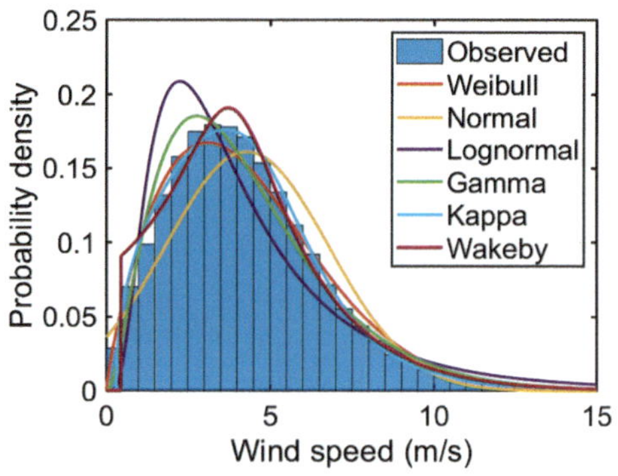

Fig. 7.2 Fit of wind speed observations with Weibull, normal, lognormal, gamma, Kappa, and Wakeby distributions

$$+ (1-\omega)\frac{k}{c}\left(\frac{x}{c}\right)^{k-1}\exp\left[-\left(\frac{x}{c}\right)^{k}\right] \tag{7.27}$$

$$F_{N-W}(x;\omega,\mu,\sigma,k,c) = \omega\frac{1}{2}\left[1+\mathrm{erf}\left(\frac{x-\mu}{\sqrt{2}\sigma}\right)\right] + (1-\omega)\left(1-\exp\left[-\left(\frac{x}{c}\right)^{k}\right]\right) \tag{7.28}$$

where ω is the mixing parameter, μ and σ are the mean and standard deviation parameters for the normal distribution mode, respectively, and k and c are the shape and scale parameters for the Weibull distribution mode, respectively.

Another example of bimodal distribution functions is the normal–normal mixture distribution (Jung & Schindler, 2017):

$$f_{N-N}(x;\omega,\mu_1,\sigma_1,\mu_2,\sigma_2) = \omega\frac{1}{\sqrt{2\pi}\sigma_1}\exp\left[-\frac{(x-\mu_1)^2}{2\sigma_1^2}\right] + (1-\omega)\frac{1}{\sqrt{2\pi}\sigma_2}\exp\left[-\frac{(x-\mu_2)^2}{2\sigma_2^2}\right] \tag{7.29}$$

$$F_{N-N}(x;\omega,\mu_1,\sigma_1,\mu_2,\sigma_2) = \omega\frac{1}{2}\left[1+\mathrm{erf}\left(\frac{x-\mu_1}{\sqrt{2}\sigma_1}\right)\right] + (1-\omega)\frac{1}{2}\left[1+\mathrm{erf}\left(\frac{x-\mu_2}{\sqrt{2}\sigma_2}\right)\right] \tag{7.30}$$

where μ_1 and σ_1 are the mean and standard deviation for the first normal distribution mode, respectively, and μ_2 and σ_2 are the mean and standard deviation for the second normal distribution mode, respectively.

Another bimodal distribution, the Weibull–Weibull mixture distribution, is expressed as (Akpinar & Akpinar, 2009):

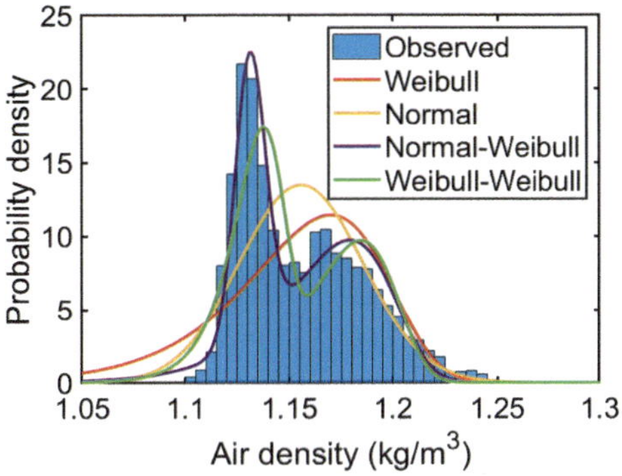

Fig. 7.3 Fit of air density observations with Weibull, normal, normal–Weibull, and Weibull–Weibull distributions

$$f_{W-W}(x;\omega,k_1,c_1,k_2,c_2)=\omega\frac{k_1}{c_1}\left(\frac{x}{c_1}\right)^{k_1-1}\exp\left[-\left(\frac{x}{c_1}\right)^{k_1}\right]+(1-\omega)\frac{k_2}{c_2}\left(\frac{x}{c_2}\right)^{k_2-1}\exp\left[-\left(\frac{x}{c_2}\right)^{k_2}\right] \tag{7.31}$$

$$F_{W-W}(x;\omega,k_1,c_1,k_2,c_2)=\omega\left(1-\exp\left[-\left(\frac{x}{c_1}\right)^{k_1}\right]\right)+(1-\omega)\left(1-\exp\left[-\left(\frac{x}{c_2}\right)^{k_2}\right]\right) \tag{7.32}$$

where k_1 and c_1 are the shape and scale parameters for the first Weibull distribution mode, respectively, and k_2 and c_2 are the shape and scale parameters for the second Weibull distribution mode, respectively.

Figure 7.3 shows the fit of air density observations with Weibull, normal, normal–Weibull, and Weibull–Weibull distributions. The air density was observed by a microwave radiometer at a height of 120 m at an urban site for a period of two years.

7.2 Long-Term Mean Wind Profiles

7.2.1 Vertical Profiles of Long-Term Mean Wind Speed

Models for vertical profiles of typically averaged (e.g., 10-min, 30-min, and 1-h mean) wind speed discussed in Chap. 4 cannot be directly applied to vertical profiles of long-term mean (e.g., annual mean) wind speed, which are of critical importance to wind resource assessment. Specifically, using a mean stability parameter, such as the mean Obukhov length, to calculate wind profiles within the MOST framework can introduce systematic biases. This occurs due to the nonlinear nature of the stability correction function, i.e., $\overline{\psi_m(z/L)} \neq \psi_m(z/\overline{L})$ (Sect. 4.2.1), and the characteristics of the stability parameter distribution.

To address this issue, Kelly and Gryning (2010) proposed a model of the climatological mean wind profile in the surface layer, as follows:

$$\overline{\left(\frac{kU}{u_*}\right)} = \ln\left(\frac{z}{\overline{z_0}}\right) - \left[\overline{\psi_m(z)} - \overline{\psi_0}\right] \tag{7.33}$$

where $\overline{\psi_m(z)}$ is the long-term stability correction based on a given stability distribution $f(1/L)$, and $\overline{\psi_0}$ is a constant which originates from the integration of the small "surface correction" $\psi_m(z_0/L)$.

The extended version of this model beyond the surface layer is given as:

$$\begin{aligned}\overline{\left(\frac{kU}{u_*}\right)} &= \ln\left(\frac{z}{\overline{z_0}}\right) - \left(1 - \frac{z}{\overline{h}}\right)\overline{\psi_m(z)} - \frac{1}{\overline{h}}\int_z^{z_0} \overline{\psi_m(z)}dz \\ &+ \frac{\overline{h}}{2\overline{L_{MBL}}}\left[1 - \left(1 - \frac{z}{\overline{h}}\right)^2\right]\end{aligned} \tag{7.34}$$

A realistic and general representation of $f(1/L)$ is needed to determine the climatological mean wind profile. Kelly and Gryning (2010) expected a probability distribution of the form $f(1/L) \propto \exp\left(-|1/L|^{2/3}\right)$ for typical ABL regimes affected by surface heat flux:

$$f(1/L) = n_\pm \frac{C_\pm}{\sigma_\pm} \frac{\exp\left[-(C_\pm|1/L|/\sigma_\pm)^{2/3}\right]}{\Gamma(5/2)} \tag{7.35}$$

$$\sigma_\pm = \frac{g}{\theta_{v0}} \frac{\sqrt{\overline{\left(H - \overline{H}_\pm\right)^2}_\pm}}{\overline{u_*^3}} \tag{7.36}$$

where subscripts $+$ and $-$ denote the stable and unstable sides of the distribution, respectively, n_+ and n_- are the fraction of occurrence of stable and unstable conditions, which together with the normalization coefficient $C_\pm/\left[\sigma_\pm\Gamma(5/2)\right]$ ensure that $\int_{-\infty}^{\infty} f(1/L) = 1$, $C_+ = 1$ and $C_- = 3$ are constants that reflect the different physical characteristics of stable and unstable flow regimes, respectively, and $H = \overline{w'\theta_v'}$ is the kinematic heat flux.

7.2.2 Vertical Profiles of Weibull Shape Parameter

Only knowing the mean wind speed is not enough for wind resource assessment. Higher moments and tail behaviors of the wind speed distribution also influence the wind resources. Assuming that the wind speed distribution follows the Weibull distribution, then a 10% increase in Weibull shape parameter k can lead to about $\pm$ 5% change in wind power density (Kelly et al., 2014). Therefore, to predict the wind power density at the turbine hub height using wind speed distribution observed at a lower height, not only a model for the vertical profile of mean wind speed but also a model for the vertical profile of k is required.

One of the earliest models for such a purpose was established by Justus and Mikhail (1976), which predicts that k increases with altitude, as follows:

$$k = k_s \frac{1 - 0.881 \ln(z_s/10)}{1 - 0.881 \ln(z/10)} \tag{7.37}$$

where k_s is the shape parameter at a height near the surface z_s (usually $z_s = 10\,\text{m}$), and k is the shape parameter at height z. Note that the model by Justus and Mikhail (1976) is only applicable to heights below 100 m. Above 100 m, k was observed to reach a maximum at the so-called "reversal height" and decrease with height above it.

To account for this vertical variation, Wieringa (1989) proposed the following model:

$$k = k_s + c_k(z - z_s)\exp\left(-\frac{z - z_s}{z_r - z_s}\right) \tag{7.38}$$

where $c_k \cong 0.02\,\text{m}^{-1}$ is a constant, and z_r is the reversal height where k attains a maximum.

To approximate z_r, Troen and Petersen (1989) used the resistance law in combination with the logarithmic wind profile to establish a simple model, as follows:

$$z_r = \alpha_r Ro^{\beta} z_0 = \alpha_r \left(\frac{U_g}{f}\right)^{\beta} z_0^{1-\beta} \tag{7.39}$$

where α_r is a constant of proportionality taken as 0.002 (Troen & Petersen, 1989) or 0.003 (Kelly et al., 2014), and β is a dimensionless parameter taken as 0.9 (Kelly et al., 2014).

Kelly et al. (2014) updated the parameterization of Wieringa (1989) using the following form without the arbitrary constant c_k and with the proper dimensions:

$$k = k_s \frac{1 + (z/z_r)\exp(-z/z_r)}{1 + (z_s/z_r)\exp(-z_s/z_r)} \tag{7.40}$$

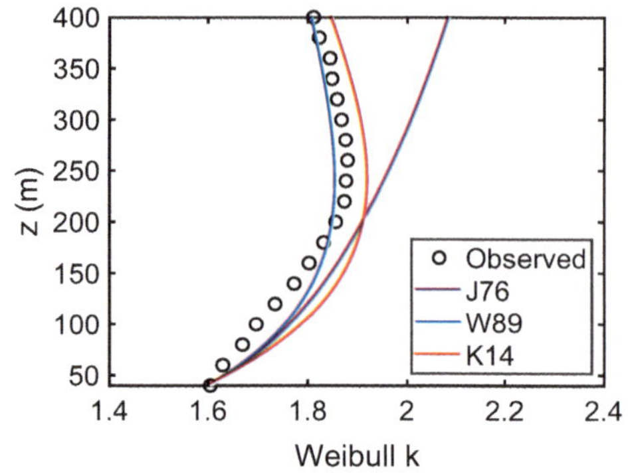

Fig. 7.4 Vertical profile of Weibull shape parameter

Furthermore, Gryning et al. (2014) added a term to the form proposed in Wieringa (1989) to allow k to approach a given value k_t at z_t in the upper boundary layer:

$$k = k_s + c\frac{z - z_s}{z_r - z_s}\exp\left(-\frac{z - z_s}{z_r - z_s}\right) - (k_s - k_t)\exp\left(-\frac{z_t - z_s}{z - z_s}\right) \tag{7.41}$$

where typical values of the parameterization constant c is $c \cong 1$, and $k_t \cong 2$ at $z_t \cong 800\,\mathrm{m}$.

Figure 7.4 shows the vertical profile of Weibull shape parameter, which was calculated from wind speed observations by a sodar at a coastal site for a period of one year. The shape parameter profiles described by the aforementioned models are shown, where J76: Justus and Mikhail (1976); W89: Wieringa (1989); K14: Kelly et al. (2014).

7.3 Macroscale Wind Spectra

Compared to the turbulent spectrum in the ABL (Sects. 1.3 and 5.1.4), the spectrum at frequencies below the spectral gap remains less explored and understood. Figure 7.5 shows an example of long-term wind spectra, which was estimated from wind speed observations at four weather stations in Hong Kong.

A notable finding in the literature regarding the long-term wind spectrum is that the energy amplitude of the spectrum E decreases with wave number κ ($\kappa = 2\pi/\lambda$, where λ is the wavelength that can be approximated by $\lambda = \overline{u}/n$ based on Taylor's frozen turbulence hypothesis), following a pattern of approximately $\kappa^{-5/3}$ in the mesoscale range, and transitioning to κ^{-3} at smaller wave numbers. Lindborg (1999) derived a model combining these two spectral forms using structure function analysis for the wind spectrum across mesoscales to synoptic scales:

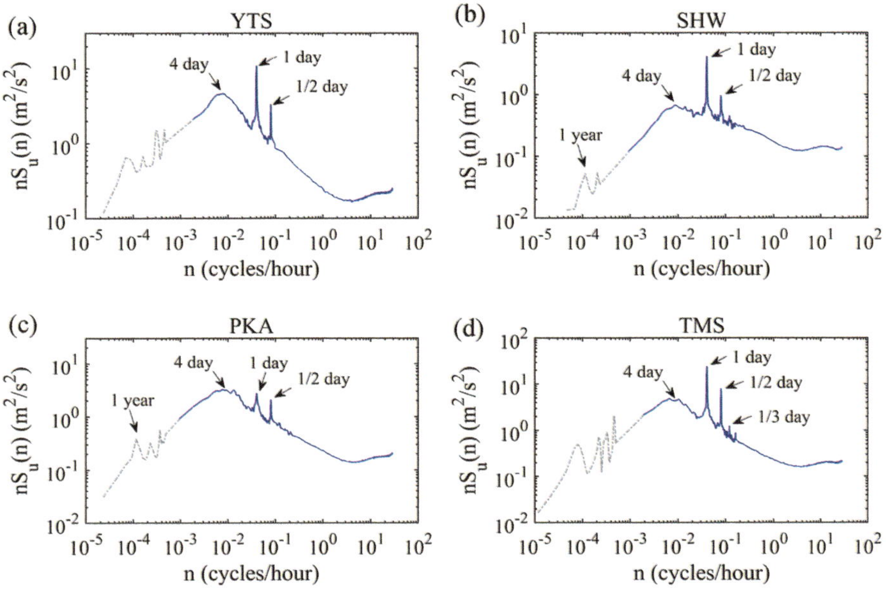

Fig. 7.5 Spectra of long-term wind speed observations at four weather stations (Lin et al., 2020)

$$E(\kappa) = d_1\kappa^{-5/3} + d_2\kappa^{-3} \tag{7.42}$$

where κ is the wavenumber in the unit of m^{-1}, $E(\kappa)$ is the power spectrum in the κ domain, and the coefficients $d_1 \approx 9.1 \times 10^{-4}\,m^{4/3}\,s^{-2}$ and $d_2 \approx 3 \times 10^{-10}\,s^{-2}$.

Larsén et al. (2013) observed similar spectral behavior in climatological long-term time series across the frequency domain, specifically for scales ranging from several days down to approximately 10 min:

$$S(n) = a_1 n^{-5/3} + a_2 n^{-3} \tag{7.43}$$

where n is the frequency, and the coefficients $a_1 = 3 \times 10^{-4}\,m^2\,s^{-8/3}$ and $a_1 = 3 \times 10^{-11}\,m^2\,s^{-4}$.

The κ^{-3} behavior is generally interpreted as an enstrophy cascade, where mean square vorticity flows from large scales to smaller scales. In contrast, the $\kappa^{-5/3}$ behavior is associated with an energy cascade from smaller scales to larger scales. Geostrophic turbulence related to baroclinic instability serves as the source of enstrophy injection at the larger scale end of the κ^{-3} region, while buoyancy forces, such as gravity waves or convective cloud systems, may inject energy at the smaller scale end of the $\kappa^{-5/3}$ range.

7.4 Climate Change

Climate change, primarily driven by greenhouse gas emissions from fossil fuel combustion, is expected to significantly alter Earth's physical environment and ecosystems, impacting human health and livelihoods. As global temperatures rise and atmospheric patterns shift, the spatial and temporal distributions of winds are likely to change substantially under climate change.

7.4.1 Global Climate Models

Research aimed at quantifying the impacts of climate change predominantly relies on global climate models (GCMs). These models provide three-dimensional wind data across the globe, typically with horizontal resolutions ranging from 100 to 250 km, and temporal coverage until 2100.

GCMs utilize mathematical equations to simulate the processes and interactions that govern Earth's climate, encompassing the atmosphere, oceans, land, and ice (Pasini, 2005). Key equations include the Navier–Stokes equations of fluid motion, the first law of thermodynamics, the Clausius–Clapeyron equation, and the Stefan–Boltzmann Law.

To project future climate changes, GCMs incorporate external factors, or "forcings", that influence the amount of solar energy absorbed by the Earth or retained by the atmosphere. These forcings encompass variations in solar output, long-lived greenhouse gases (such as CO_2, CH_4, N_2O, and halocarbons), and aerosols emitted from fossil fuel combustion, wildfires, and volcanic activity (Stocker, 2011). Future "emission scenarios" outline plausible pathways for radiative forcing, based on anticipated developments in technology, energy use, and land management (IPCC, 2021).

The impacts of climate change on wind patterns can be evaluated using global climate models (GCMs) from the Coupled Model Intercomparison Project Phase 6 (CMIP6) (Eyring et al., 2016), the latest iteration following Phase 5 (CMIP5). Initiated in 2015 under the World Climate Research Programme (WCRP), CMIP6 is a collaborative international effort aimed at improving understanding of climate change. It includes simulations from approximately 100 GCMs developed by 49 modeling groups worldwide and provides the foundation for the Sixth Assessment Report (AR6) of the United Nations Intergovernmental Panel on Climate Change (IPCC).

To project future developments in anthropogenic climate drivers, such as greenhouse gases and land use, CMIP6 features the Scenario Model Intercomparison Project (ScenarioMIP) (O'Neill et al., 2016). This new integrated framework merges expected radiative forcing with Shared Socioeconomic Pathways (SSPs), which outline different future scenarios based on urbanization, population growth, and economic expansion (Riahi et al., 2017). The five SSP narratives range from sustainable development (SSP1) to a high-energy, fossil-fuel-dependent economy (SSP5) (O'Neill et al., 2017). In ScenarioMIP, each SSP is paired with a projected radiative forcing level in W/m^2 for 2100, resulting in scenarios such as SSPx-y. Figure 7.6 shows an illustrative diagram of the SSP scenarios.

In climate studies, the intermediate scenario SSP2-4.5, which follows historical socioeconomic trends with moderate radiative forcing of 4.5 W/m^2, as well as the more extreme scenario SSP5-8.5, characterized by high fossil fuel use and a radiative forcing of 8.5 W/m^2, are usually selected to represent the future climate pathways. These scenarios suggest global temperature increases of 2.1–3.5 °C and 3.3–5.7 °C, respectively, by the end of the twenty-first century. Other scenarios, such as SSP1-2.6 and SSP3-7.0, are also considered. The SSPs are related to the Representative Concentration Pathways (RCPs) from CMIP5, with SSP2-4.5 and SSP5-8.5 being similar to RCP4.5 and RCP8.5, respectively, though not identical. Figure 7.7 depicts the SSP and RCP concentrations for CO_2.

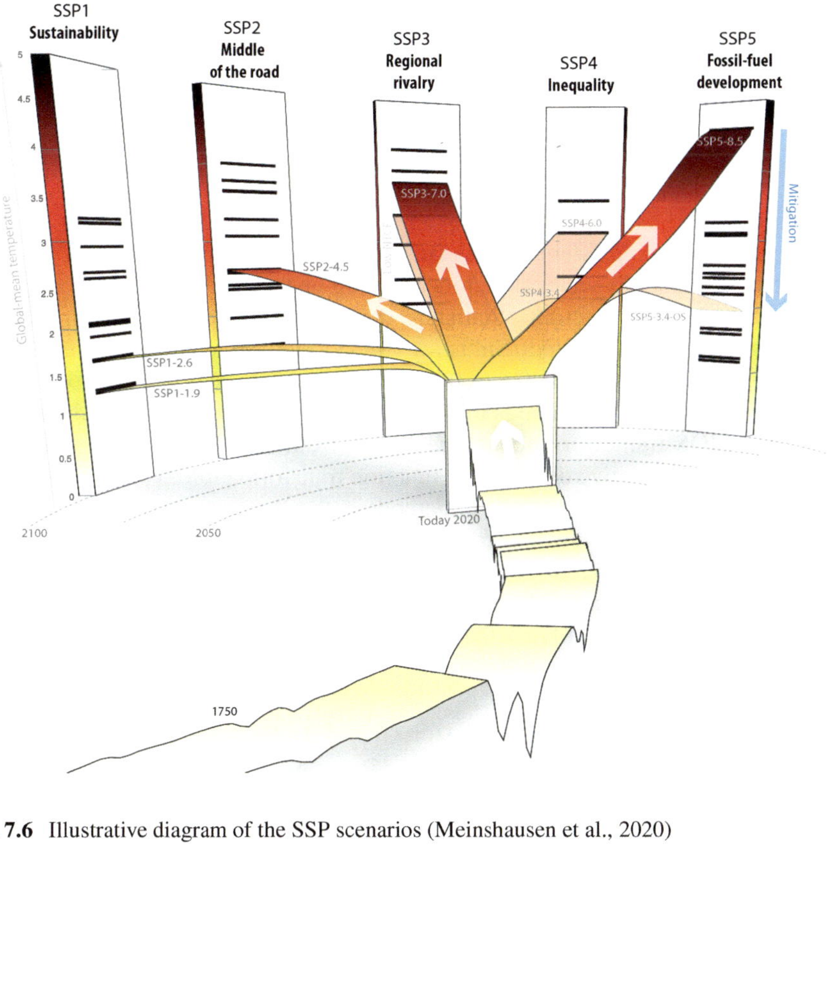

Fig. 7.6 Illustrative diagram of the SSP scenarios (Meinshausen et al., 2020)

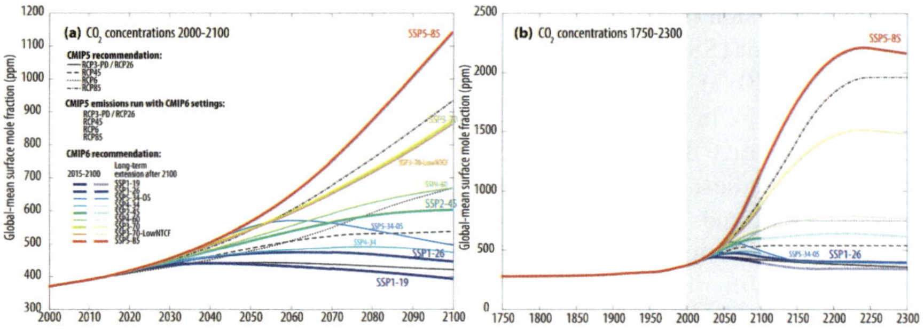

Fig. 7.7 SSP and RCP concentrations for CO_2 (Meinshausen et al., 2020)

The CMIP6 model data are accessible from several portals, such as the portal of the German Climate Computing Center (Deutsches Klimarechenzentrum, DKRZ) via https://esgf-data.dkrz.de/projects/cmip6-dkrz/. To enhance spatial resolution and enable regional assessments, various downscaling methods have been developed, including statistical and dynamical downscaling, which will be introduced in the following sections.

7.4.2 Statistical Downscaling

Statistical downscaling establishes a statistical relationship between local variables (predictands) and large-scale predictors based on historical data, which is then used to downscale future projections. Statistical downscaling methods, including Perfect Prognosis (PP), Model Output Statistics (MOS), and Weather Generators (WG), are employed to translate wind data from global scales to regional and local levels. Among these, the MOS method is the most commonly used for downscaling winds. Within the MOS framework, techniques range from simple additive correction to more complex cumulative distribution function transformation.

One of the simplest downscaling methods is the linear approach, which can be implemented through additive correction or additive scaling. The additive correction method predicts small-scale wind speeds in the forecast period by adding the mean difference between calibration period wind speeds, while the additive scaling method does this by multiplying the mean ratio of calibration period wind speeds. The additive correction method is:

$$y_f = x_f + (\overline{y_c} - \overline{x_c}) \tag{7.44}$$

where x_f and y_f are the small-scale wind speed and large-scale wind speed in the forecast period, respectively, and x_c and y_c are those in the calibration period. The overbar denotes the mean over the period.

The additive scaling method is:

$$y_f = x_f\left(\frac{\overline{y_c}}{\overline{x_c}}\right) \tag{7.45}$$

Both additive correction and additive scaling methods operate under the assumption that the discrepancies between observed and modeled wind speeds remain constant over time, and that higher-order statistics, such as variance and skewness, are identical. However, differences in these statistical properties can exist, necessitating adjustments during the downscaling process. Variance correction can be applied to account for these discrepancies by incorporating differences in mean wind speeds and standard deviations:

$$y_f = x_f + (\overline{y_c} - \overline{x_c}) + \left(x_f - \overline{x_f}\right)\left(\frac{\sqrt{\mathrm{E}\left[(y_c - \overline{y_c})^2\right]}}{\sqrt{\mathrm{E}\left[(x_c - \overline{x_c})^2\right]}}\right) \tag{7.46}$$

A more advanced technique is the quantile mapping method, which aligns the quantiles of large-scale wind speeds with those of local wind speeds. This involves estimating cumulative distribution functions (CDFs) for the calibration period, which can be derived using parametric or non-parametric statistical models. The quantile mapping method is:

$$y_f = F_{yc}^{-1}\left[F_{xc}\left(x_f\right)\right] \tag{7.47}$$

where F_{xc} and F_{yc} represent the CDFs of x_c and y_c in the calibration period, respectively. In the case of extreme values outside the calibration range, constant extrapolation is utilized based on the highest and lowest quantiles.

An extension of the quantile mapping method is the cumulative distribution function transform method. Unlike quantile mapping, which only maps CDFs from the calibration period, the cumulative distribution function transform method also considers the CDFs of large-scale wind speeds in the forecast period. The CDF for the forecast period is derived, and subsequent wind speeds are calculated through quantile mapping:

$$F_{yf}(x) = F_{yc}\left(F_{xc}^{-1}\left[F_{xf}(x)\right]\right) \tag{7.48}$$

$$y_f = F_{yf}^{-1}\left[F_{xf}\left(x_f\right)\right] \tag{7.49}$$

where F_{xf} and F_{yf} represent the CDFs of x_f and y_f in the forecast period, respectively.

Both quantile mapping and cumulative distribution function transform methods can utilize empirical CDFs with linearly interpolated quantiles and constant corrections for out-of-range values. Alternatively, parametric models, such as the Weibull distribution, can also be employed for estimating the necessary CDFs.

7.4.3 Dynamical Downscaling

In contrast to statistical downscaling, dynamical downscaling employs regional climate models (RCMs) to simulate detailed meteorological fields, using boundary conditions derived from GCM. The concept of using high-resolution meteorological models to refine coarse GCM data was initially proposed through large ensembles of short simulations lasting 3–5 days, following established weather prediction methodologies. The first RCM simulations for climate modeling were designed to run continuously beyond the synoptic weather scale, using meteorological lateral boundary conditions derived from either observational data or GCM outputs. This marked a significant advancement in regional climate modeling, demonstrating that RCM performance remains stable after a brief spin-up period.

Today, various RCM systems are available, developed from mesoscale and weather forecasting models or as regional adaptations of global models. These RCMs are utilized by numerous institutions globally, proving to be versatile tools for a diverse community engaged in a range of applications, including regional process studies, sensitivity analyses, paleoclimate investigations, and future climate projections across all land areas. This adaptability is particularly beneficial for scientists in developing regions, allowing them to conduct advanced research without the extensive infrastructure typically required for high-quality GCMs.

To generate a comprehensive set of high-resolution regional climate projections for all major continental areas, the WCRP has launched a coordinated initiative known as the Coordinated Regional Climate Downscaling Experiment (CORDEX). Figure 7.8 shows examples of CORDEX model domains.

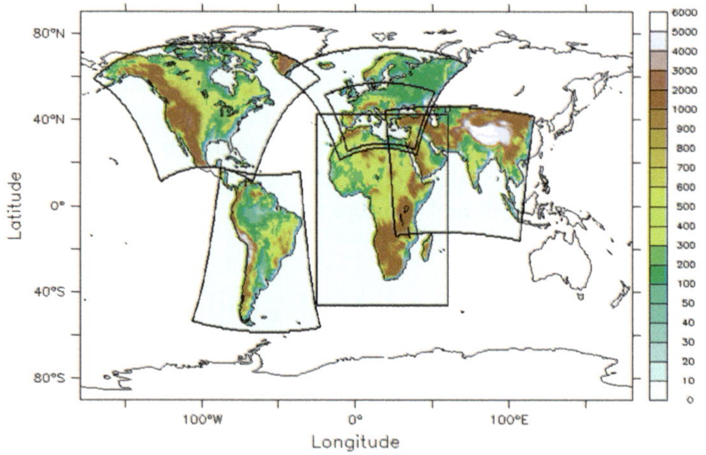

Fig. 7.8 Examples of Coordinated Regional Climate Downscaling Experiment model domains (Jacob et al., 2012)

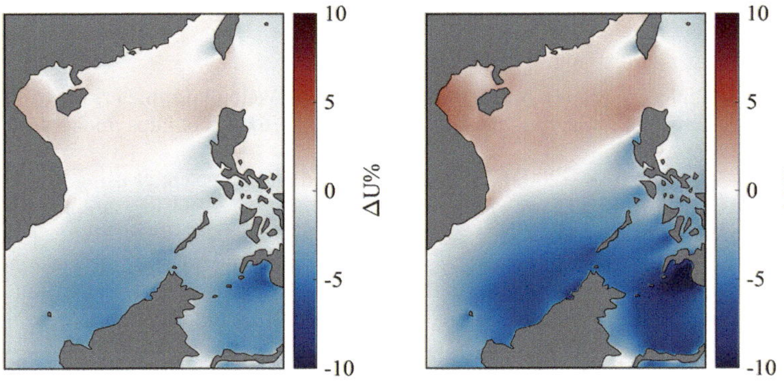

Fig. 7.9 Percentage change in wind speed at 100-m height over the South China Sea for 2046–2065 (left) and 2081–2100 (right) relative to 1985–2004 (He et al., 2023)

7.4.4 Climate Change Impacts on Winds

Climate change is anticipated to weaken the westerly wind belts globally, primarily because polar regions are warming more rapidly than tropical areas. This differential warming will reduce the meridional temperature gradient, which is a key driver of global westerlies. Furthermore, the diminishing temperature gradient may result in a poleward shift of storm tracks. Beyond these general effects, climate change can also cause changes in regional or local weather patterns, such as predominant wind directions and storm behaviors, resulting in significant variations in the wind climate at specific locations.

As an example, Fig. 7.9 illustrates the spatial distribution of percentage changes in mean wind speed at 100-m height over the South China Sea (SCS), for the periods 2046–2065 and 2081–2100 relative to 1985–2004 and under the RCP8.5 scenario. The climate change signal was quantified using a regional climate model RegCM4.7. Under climate change, mean wind speed is projected to increase in the northern SCS while decreasing in the southern SCS. Specifically, the anticipated increase is about 1–3% during 2046–2065 and 2–5% in 2081–2100. Conversely, the decrease is estimated to be around 1–5% for 2046–2065 and 2–10% for 2081–2100. These variations in wind speed are likely linked to the northward shift and increased variability of the western Pacific subtropical high (WPSH) in the future climate. This movement of the WPSH is expected to influence the East Asian monsoon systems, subsequently affecting wind speed in the region.

References

Akdağ, S. A., & Dinler, A. (2009). A new method to estimate Weibull parameters for wind energy applications. *Energy Conversion and Management, 50*, 1761–1766. https://doi.org/10.1016/j.enconman.2009.03.020

Akpinar, S., & Akpinar, E. K. (2009). Estimation of wind energy potential using finite mixture distribution models. *Energy Conversion and Management, 50*, 877–884. https://doi.org/10.1016/j.enconman.2009.01.007

Alavi, O., Sedaghat, A., & Mostafaeipour, A. (2016). Sensitivity analysis of different wind speed distribution models with actual and truncated wind data: A case study for Kerman, Iran. *Energy Conversion and Management, 120*, 51–61. https://doi.org/10.1016/j.enconman.2016.04.078

Alexandre, P., Rocha, C., Sousa, R. C. D., Andrade, C. F. D., & Eugênia, M. (2012). Comparison of seven numerical methods for determining Weibull parameters for wind energy generation in the northeast region of Brazil. *Applied Energy, 89*, 395–400. https://doi.org/10.1016/j.apenergy.2011.08.003

Amirinia, G., Kamranzad, B., & Ma, S. (2017). Wind and wave energy potential in southern Caspian Sea using uncertainty analysis. *Energy, 120*, 332–345. https://doi.org/10.1016/j.energy.2016.11.088

Carta, J. A., Ramírez, P., & Velázquez, S. (2009). A review of wind speed probability distributions used in wind energy analysis. Case studies in the Canary Islands. *Renewable and Sustainable Energy Reviews, 13*, 933–955. https://doi.org/10.1016/j.rser.2008.05.005

Chang, T. P. (2011a). Estimation of wind energy potential using different probability density functions. *Applied Energy, 88*, 1848–1856. https://doi.org/10.1016/j.apenergy.2010.11.010

Chang, T. P. (2011b). Performance comparison of six numerical methods in estimating Weibull parameters for wind energy application. *Applied Energy, 88*, 272–282. https://doi.org/10.1016/j.apenergy.2010.06.018

Eyring, V., Bony, S., Meehl, G. A., Senior, C. A., Stevens, B., Stouffer, R. J., & Taylor, K. E. (2016). Overview of the Coupled Model Intercomparison Project Phase 6 (CMIP6) experimental design and organization. *Geoscientific Model Development, 9*(5), 1937–1958. https://doi.org/10.5194/gmd-9-1937-2016

Gil Ruiz, S. A., Barriga, J. E. C., & Martínez, J. A. (2021). Wind power assessment in the Caribbean region of Colombia, using ten-minute wind observations and ERA5 data. *Renewable Energy, 172*, 158–176. https://doi.org/10.1016/j.renene.2021.03.033

Gryning, S. E., Batchvarova, E., Floors, R., Peña, A., Brümmer, B., Hahmann, A. N., & Mikkelsen, T. (2014). Long-term profiles of wind and Weibull distribution parameters up to 600 m in a rural coastal and an inland suburban area. *Boundary-Layer Meteorology, 150*, 167–184. https://doi.org/10.1007/s10546-013-9857-3

He, J. Y., Chan, P. W., Li, Q. S., & Tong, H. W. (2023). Mapping future offshore wind resources in the South China Sea under climate change by regional climate modeling. *Renewable and Sustainable Energy Reviews, 188*, 113865. https://doi.org/10.1016/j.rser.2023.113865

Hosking, J. R. M., & Wallis, J. R. (1997). *Regional frequency analysis*. Cambridge University Press.

IPCC. (2021). *Climate change 2021: The physical science basis.*

Jacob, D., Elizalde, A., Haensler, A., Hagemann, S., Kumar, P., Podzun, R., Rechid, D., Remedio, A. R., Saeed, F., Sieck, K., Teichmann, C., & Wilhelm, C. (2012). Assessing the transferability of the regional climate model REMO to different COordinated Regional Climate Downscaling EXperiment (CORDEX) regions. *Atmosphere, 3*(1), 181–199. https://www.mdpi.com/2073-4433/3/1/181

Jung, C., & Schindler, D. (2017). Global comparison of the goodness-of-fit of wind speed distributions. *Energy Conversion and Management, 133*, 216–234. https://doi.org/10.1016/j.enconman.2016.12.006

Justus, C. G., Hargraves, W. R., Mikhail, A., & Graber, D. (1978). Methods for estimating wind speed frequency distributions. *Journal of Applied Meteorology, 17*, 350–353.

Justus, C. G., & Mikhail, A. (1976). Height variation of wind speed and wind distributions statistics. *Geophysical Research Letters, 3*, 261–264. https://doi.org/10.1029/GL003i005p00261

Kelly, M., & Gryning, S. E. (2010). Long-term mean wind profiles based on similarity theory. *Boundary-Layer Meteorology, 136*, 377–390. https://doi.org/10.1007/s10546-010-9509-9

Kelly, M., Troen, I., & Jørgensen, H. E. (2014). Weibull-k revisited: "Tall" profiles and height variation of wind statistics. *Boundary-Layer Meteorology, 152*, 107–124. https://doi.org/10.1007/s10546-014-9915-5

Larsén, X. G., Vincent, C., & Larsen, S. (2013). Spectral structure of mesoscale winds over the water. *Quarterly Journal of the Royal Meteorological Society, 139*, 685–700. https://doi.org/10.1002/qj.2003

Li, Q. S., Shu, Z. R., & Chen, F. B. (2016). Performance assessment of tall building-integrated wind turbines for power generation. *Applied Energy, 165*, 777–788. https://doi.org/10.1016/j.apenergy.2015.12.114

Lin, H. B., Fu, J. Y., Shu, Z. R., Li, Q. S., Chan, P. W., & He, Y. C. (2020). Spectral characteristics of surface atmosphere in range of macroscale to microscale at Hong Kong. *Journal of Wind Engineering & Industrial Aerodynamics,* 104446. https://doi.org/10.1016/j.jweia.2020.104446

Lindborg, E. (1999). Can the atmospheric kinetic energy spectrum be explained by two-dimensional turbulence? *Journal of Fluid Mechanics, 388*, 259–288. https://doi.org/10.1017/S0022112099004851

Mary, A. A. (2021). Geospatial mapping of micro-wind energy for district electrification in Ghana. *Energy, 225*, 120217. https://doi.org/10.1016/j.energy.2021.120217

Meinshausen, M., Nicholls, Z. R. J., Lewis, J., Gidden, M. J., Vogel, E., Freund, M., Beyerle, U., Gessner, C., Nauels, A., Bauer, N., Canadell, J. G., Daniel, J. S., John, A., Krummel, P. B., Luderer, G., Meinshausen, N., Montzka, S. A., Rayner, P. J., Reimann, S., … Wang, R. H. J. (2020). The shared socio-economic pathway (SSP) greenhouse gas concentrations and their extensions to 2500. *Geoscientific Model Developement, 13*(8), 3571–3605. https://doi.org/10.5194/gmd-13-3571-2020

Mohammadi, K., Alavi, O., Mostafaeipour, A., Goudarzi, N., & Jalilvand, M. (2016). Assessing different parameters estimation methods of Weibull distribution to compute wind power density. *Energy Conversion and Management, 108*, 322–335. https://doi.org/10.1016/j.enconman.2015.11.015

Morgan, E. C., Lackner, M., Vogel, R. M., & Baise, L. G. (2011). Probability distributions for offshore wind speeds. *Energy Conversion and Management, 52*, 15–26. https://doi.org/10.1016/j.enconman.2010.06.015

O'Neill, B. C., Kriegler, E., Ebi, K. L., Kemp-Benedict, E., Riahi, K., Rothman, D. S., van Ruijven, B. J., van Vuuren, D. P., Birkmann, J., Kok, K., Levy, M., & Solecki, W. (2017). The roads ahead: Narratives for shared socioeconomic pathways describing world futures in the 21st century. *Global Environmental Change, 42*, 169–180. https://doi.org/10.1016/j.gloenvcha.2015.01.004

O'Neill, B. C., Tebaldi, C., van Vuuren, D. P., Eyring, V., Friedlingstein, P., Hurtt, G., Knutti, R., Kriegler, E., Lamarque, J.-F., Lowe, J., Meehl, G. A., Moss, R., Riahi, K., & Sanderson, B. M. (2016). The scenario model intercomparison project (ScenarioMIP) for CMIP6. *Geoscientific Model Development, 9*(9), 3461–3482. https://doi.org/10.5194/gmd-9-3461-2016

Pasini, A. (2005). *From observations to simulations: A conceptual introduction to weather and climate modelling.* World Scientific.

Riahi, K., van Vuuren, D. P., Kriegler, E., Edmonds, J., O'Neill, B. C., Fujimori, S., Bauer, N., Calvin, K., Dellink, R., Fricko, O., Lutz, W., Popp, A., Cuaresma, J. C., Kc, S., Leimbach, M., Jiang, L., Kram, T., Rao, S., Emmerling, J., … Tavoni, M. (2017). The Shared Socioeconomic Pathways and their energy, land use, and greenhouse gas emissions implications: An overview. *Global Environmental Change, 42*, 153–168. https://doi.org/10.1016/j.gloenvcha.2016.05.009

Samal, R. K. (2021). Assessment of wind energy potential using reanalysis data: A comparison with mast measurements. *Journal of Cleaner Production, 313*, 127933. https://doi.org/10.1016/j.jclepro.2021.127933

Smirnov, N. (1948). Table for estimating the goodness of fit of empirical distributions. *The Annals of Mathematical Statistics, 19*, 279–281. https://doi.org/10.1214/aoms/1177730256
Stocker, T. (2011). *Introduction to climate modelling* (1st ed.). Springer. https://doi.org/10.1007/978-3-642-00773-6
Troen, I., & Petersen, E. L. (1989). *European wind atlas.*
Wais, P. (2017). A review of Weibull functions in wind sector. *Renewable and Sustainable Energy Reviews, 70*, 1099–1107. https://doi.org/10.1016/j.rser.2016.12.014
Wieringa, J. (1989). Shapes of annual frequency distributions of wind speed observed on high meteorological masts. *Boundary-Layer Meteorology, 47*, 85–110. https://doi.org/10.1007/BF00122324
Wu, J., Wang, J., & Chi, D. (2013). Wind energy potential assessment for the site of Inner Mongolia in China. *Renewable and Sustainable Energy Reviews, 21*, 215–228. https://doi.org/10.1016/j.rser.2012.12.060
Yang, L., Rojas, J. I., & Montlaur, A. (2020). Advanced methodology for wind resource assessment near hydroelectric dams in complex mountainous areas. *Energy, 190*, 116487. https://doi.org/10.1016/j.energy.2019.116487

Chapter 8
Outlook

As we navigate the complexities of winds in the ABL, it becomes evident that understanding these dynamics is crucial not only for meteorology but also for a wide range of engineering applications. The ABL is a dynamic interface where atmospheric processes interact with the Earth's surface, influencing everything from weather patterns to the performance of structures and wind energy systems.

The interplay between theory and practical application is paramount. Advances in observational technologies, such as remote sensing and sophisticated modeling techniques, have enhanced our ability to characterize wind profiles and turbulence characteristics accurately. This knowledge is essential for ensuring the safety and serviceability of buildings, optimizing the design and placement of wind turbines, and mitigating the impacts of extreme weather events.

Moreover, the ongoing challenges posed by climate change necessitate a deeper understanding of how wind patterns may evolve. Engineers and meteorologists must collaborate to develop adaptive strategies that account for these changes, ensuring that buildings and infrastructure remain resilient in the face of shifting atmospheric conditions.

As we look to the future, interdisciplinary research will be vital. Bridging the gap between meteorology and engineering will facilitate the development of innovative solutions that harness wind energy more effectively, improve urban planning, and enhance our ability to predict and respond to weather-related hazards. By fostering this collaboration, we can better prepare for the complexities of the ABL and its far-reaching implications for society.

J. He and Q.-S. Li, *Winds in the Atmospheric Boundary Layer*,
https://doi.org/10.1007/978-981-95-1517-2_8

GPSR Compliance

The European Union's (EU) General Product Safety Regulation (GPSR) is a set of rules that requires consumer products to be safe and our obligations to ensure this.

If you have any concerns about our products, you can contact us on ProductSafety@springernature.com

In case Publisher is established outside the EU, the EU authorized representative is:

Springer Nature Customer Service Center GmbH
Europaplatz 3
69115 Heidelberg, Germany

Batch number: 10704307

Printed by Printforce, the Netherlands